Exercises in Process Safety

Exercises in Process Safety

Example calculations in risk and reliability.

A revision aid for the TÜV Functional Safety Engineer (SIS) examination.

KJ Kirkcaldy

Exercises in Process Safety

First published 2016.

Copyright: © 2016 KJ Kirkcaldy. All rights reserved.

ISBN 978-1-326-55765-2

CONTENTS

1.	Notation	1
2.	Dimensions	9
3.	Initiating Event Frequencies	13
4.	The ALARP Principle	18
5.	Determining Safety Integrity Level (SIL) Targets	23
6.	Reliability Techniques	33
7.	Dangerous and Safe Failures	41
8.	System Modelling	47
9.	Requirements for Hardware Fault Tolerance	52
10.	Answers	58
11.	References	79
12.	Definitions	80
13.	Index	88

ABBREVIATIONS

λ	Failure rate, the ratio of the total number of failures occurring in a given period of time
λD	failure rate of dangerous failures
λDD	failure rate of dangerous failures detected by diagnostics
λDU	failure rate of dangerous failures undetected by diagnostics
λS	failure rate of safe failures
θ	Theta (MTBF)
1oo1	1 out of 1 voting (Simplex)
1oo2	1 out of 2
ALARP	As Low As Reasonably Practicable
BPCS	Basic Process Control System
CCF	Common Cause Failure
Dangerous failure	This is a failure mode that has the potential to put the safety-related system into a hazardous or fail-to-function state
DD	Dangerous Detected
DU	Dangerous Undetected
E/E/PES	Electrical / Electronic / Programmable Electronic System
ESD	Emergency Shutdown
ESDV	Emergency Shutdown Valve
F&G	Fire and Gas
f/hr	Failures per hour
FSC	Functional Safety Capability
HFT	Hardware Fault Tolerance
IEC	International Electrotechnical Commission
IPL	Independent Protection Layer
MDT	Mean Down Time
MooN	M out of N (general case)
MTBF	Mean Time Between Failures
MTTF	Mean Time To Failure
MTTR	Mean Time To Repair
Non-SR	Non-Safety Related
P&ID	Piping and Instrumentation Diagram
pa	Per Annum (per year)
PE	Programmable Electronics

ABBREVIATIONS

PFD	Probability of Failure on Demand
PFD_{avg}	Average Probability of Failure on Demand
PFH	Probability of Failure per Hour
PSD	Process Shutdown
PSV	Pressure Safety Valve
PT	Pressure Transmitter
PTI	Proof Test Interval
RBD	Reliability Block Diagram
RRF	Risk Reduction Factor
S	Safe
Safe failure	This is a failure mode which does not have the potential to put the safety-related system in a hazardous or fail-to-function state.
SFF	Safe Failure Fraction.
SIF	Safety Instrumented Function
SIL	Safety Integrity Level.
SIS	Safety Instrumented System
SOV	Solenoid Operated Valve
STR	Spurious Trip Rate
T_p	Proof Test Interval
VPF	Value of Preventing a Statistical Fatality

Forward

This book is aimed at Functional Safety Engineers who wish to improve their understanding of risk and reliability calculations.

Examples have been created in the calculation of various risk and reliability scenarios. Answers are also provided to enable the student to confirm understanding and consolidate knowledge.

This book may be a useful revision aid to those studying for the TÜV Functional Safety Engineer (Safety Instrumented System) examination.

This book should be used alongside recommended pre-reading:

Functional Safety in the Process Industry: A handbook of practical guidance in the application of IEC61511 and ANSI/ISA-84.00.01.

KJ Kirkcaldy and D Chauhan

ISBN 978-1-291-18723-6.

References to the relevant chapters in the pre-reading are provided.

Disclaimer

Whilst every effort has been made to ensure that the information provided here is correct, this book is intended as a revision aid only. Its objective is to enable the student to become more familiar with the concept of functional safety.

In all safety work, the analysis of risk and the measures taken to mitigate and protect against risk, the development of a safety case and the presentation of supporting evidence remain the responsibility of the duty holder (the risk owner).

The use of the square bracket [ ] indicates a cross-reference to a section within this book.

1. NOTATION

Introduction

Because the calculation of risk and reliability generally involves very small or very large numbers, the use of scientific or engineering notation is invaluable because it allows a convenient way of representing these inconvenient values.

Familiarity with this notation is therefore essential to being able to carry out such calculations and this section presents an introduction to, or a refresher in the use of scientific and engineering notation.

Scientific Notation

Scientific notation is just an easy way of representing numbers. Starting with some large numbers, we know for example, that 10 x 10 = 100.

We also know that 10 x 10 is 10^2 (we say 10-squared, or 10 to the power 2). Therefore as $10^2 = 100$, we can represent the number 100 in either way: 100 or 10^2. They are the same thing.

Similarly, 10 x 10 x 10 = 1000. Therefore $10^3 = 1000$ (we say 10-cubed, or 10 to the power 3).

So very large numbers can be expressed as powers of 10. For example: $1,000,000 = 10^6$ and $100,000,000 = 10^8$. So the 6 and the 8 in these examples (called the exponent) indicates the number of zeros, or the power of 10, in our number.

We can express other numbers in the same way:

Example

$700 = 7 \times 100 = 7 \times 10^2$

Or $700,000,000 = 7 \times 10^8$

Example

$8760 = 8.76 \times 10^3$

1. Notation

In this example, 8.76 is known as the characteristic and the part of the number to the right of the decimal point is called the mantissa. The mantissa is particularly useful as this indicates the number of significant digits we want to use.

In reliability and risk calculations, we would not expect to see more than 2 significant digits. In fact, some prefer just 1 significant digit.

Scientific notation is useful when calculating with large numbers.

Multiplying with Scientific Notation

Example

If I want to calculate the area of a field which is 1100 m by 5200 m, the area is given by 1100 x 5200 m².

Converting each number to scientific notation, we have:

Area   = $(1.1 \times 10^3) \times (5.2 \times 10^3)$

Changing the order, this is the same as:

Area   = $1.1 \times 5.2 \times 10^3 \times 10^3$

Now forgetting about the 10s for now, just calculate 1.1 x 5.2 = 5.72

∴ Area   = $5.72 \times 10^3 \times 10^3$

    = 5.72×10^6 m²

or 5.72 million metres squared.

So to multiply one number by another, we <u>multiply</u> the characteristics together, and <u>add</u> the exponents.

If we are dividing one number by another, we <u>divide</u> the characteristics and <u>subtract</u> the exponents.

Dividing with Scientific Notation

Example

For example, 10,000 divided by 1000 = 10

Using scientific notation $10^4 / 10^3 = 10^1$

To divide, we divide the characteristics and subtract the exponents.

Example

Now, if we divide 10 by 100 this would give us 0.1. Using scientific notation:

10 / 100 = $10^1 / 10^2$

Subtracting the exponents gives

= 10^{-1} = 0.1

So $0.1 = 10^{-1}$

similarly $0.01 = 10^{-2}$,

$0.001 = 10^{-3}$,

$0.0001 = 10^{-4}$ and so on.

Example

8000 / 100 = $8 \times 10^3 / 10^2$ = 8×10^1 = 80

Example

To divide 8,000 by 40,000:

8,000 / 40,000 = $8 \times 10^3 / 4 \times 10^4$

= $8 / 4 \times 10^{-1}$

= 2×10^{-1} (or 0.2)

Divide the characteristics and subtract the exponents.

1. Notation

Scientific notation also allows us to deal in a simple way with very small numbers. So, $0.0000000572 = 5.72 \times 10^{-8}$ and we can multiply and divide just as before.

If we need to divide 0.0000000572 by 0.00000011, it would be very inconvenient if we had to write out all of the leading zeros and it would be easy to make a mistake.

Expressed in scientific notation, the calculation becomes:

$5.72 \times 10^{-8} / 1.1 \times 10^{-7}$ = 5.2×10^{-1}

Remember to divide the characteristic and subtract the exponent.

To convert from decimal numbers less than 1, to scientific, the exponent shows how many places we have moved the decimal point to the right. For example,

If we have a value 0.0001, it is inconvenient to have the decimal point followed by leading zeros. But if we move the decimal point to the right, to remove the leading zeros, this is equivalent to multiplying by 10, so to keep the value the same we must also divide by 10.

Therefore 0.0001 = 0.001 $\times 10^{-1}$
              = 0.01 $\times 10^{-2}$
              = 0.1 $\times 10^{-3}$
              = 1 $\times 10^{-4}$
              = 10^{-4}

$\underset{0.0001}{\overset{1\ 2\ 3\ 4}{\frown\frown\frown}}$ = $1.0 \times 10^{-4} = 10^{-4}$

If the exponent is negative, the number is less than 1.

When we move the decimal point one place to the left, we are effectively dividing by 10, so to keep the number the same, we must also multiply by 10, e.g.

8 7 6 0   = $8 7 6 . 0 \times 10^1$

               = $8 7 . 6 0 \times 10^2$

               = $8 . 7 6 0 \times 10^3$

To convert from decimal numbers greater than 1, to scientific, the exponent is how many places we have moved the decimal point to the left:

8 7 6 0   =   8 . 7 6 0 x 10^3

Engineering Notation

Engineering notation gives a more convenient way of representing the exponent of a scientific number. In engineering notation, the exponent part of the number, e.g. x 10^{-3} is written E-03.

So, 8.76 x 10^3 = 8.76E+03

8.76 x 10^{-3} = 8.76E-03.

All the same rules apply, the 8.76 is the characteristic, the .76 is the mantissa and the -03 is the exponent.

We usually express engineering numbers with one digit to the left of the decimal point, so 220 would normally be written as 2.2E+02, and 2000 would be written as 2.0E+03.

The numbers to the right of the decimal point, the mantissa, shows the number of significant digits. Usually 1 or 2 decimal places is adequate.

Example

Express 317859 in engineering notation, to two decimal places:
317859   =   3.18E+05

Express 317859 in engineering notation, to one decimal place:
317859   =   3.2E+05

Adding and Subtracting

We can add and subtract numbers in scientific or engineering notation very simply but first the exponents must be arranged to be the same.

Example

8.76E-04 + 3.2E-03	=	8.76E-04 + 32.0E-04
	=	40.76E-04
	=	4.076E-03

Example

3.2E-03 - 8.76E-04	=	32.0E-04 - 8.76E-04
	=	23.24E-04
	=	2.324E-03

Use of Calculators

The above examples are just so that you can familiarise yourself with scientific or engineering notation and if you can follow the calculations by hand then you should have a good understanding.

Scientific calculators are equipped with an EXP (or EE, or E) button to use engineering notation. Try the above examples again using your scientific calculator.

The following pages have some example calculations which you can try either by hand or by using a scientific calculator.

1. Notation

Example 1: express the following in scientific notation to two decimal places:

a) 120
b) 10,400
c) 0.0005
d) 1
e) 0.0000000605
f) 10 million
g) 53 million
h) 0.00003
i) 2
j) 0.00000012
k) 0.0010000

Example 2: express the following scientific numbers in decimals:

a) 3×10^{-1}
b) 2×10^{-5}
c) 0.009×10^{2}
d) 100×10^{-6}
e) 300×10^{3}
f) 0.5×10^{5}
g) 3000×10^{-3}
h) 0.0005×10^{-4}
i) 1000×10^{-9}
j) 3×10^{-0}
k) 3×10^{0}

1. Notation

Example 3: express the following in engineering notation to two decimal places:

a) 40
b) 54,001
c) 0.0003
d) 1
e) 0.0000060005
f) 101 million
g) 350 million
h) 0.000013
i) 2
j) 0.0000001002
k) 0.0010000

Example 4: express the following engineering values as decimals:

a) 5.00E-01
b) 2.0E-05
c) 9.0E+02
d) 100E-06
e) 300E+03
f) 0.5E-05
g) 5000E-03
h) 0.0004E-04
i) 1000E-09
j) 1.0E+00
k) 3.00E-00

2. DIMENSIONS

We can use whatever units we find convenient for our calculations provided any equations we use are dimensionally consistent. That is, the dimensions (units) have to be the same on both sides of the equation.

If we have "/year" on one side of the equation, we must have "/year" on the other. It is important to always check this because for the equation to be correct, both sides must balance.

Usually, initiating event frequencies are expressed as /year, or /hour so we must be able to convert between the two.

To convert a frequency from /hr to /yr we must multiply by the number of hours in a year.

Dimensionally:

$$\text{Frequency} \quad \left[\frac{1}{hr}\right] \times \left[\frac{hr}{yr}\right] = \left[\frac{1}{yr}\right]$$

Cancelling out the [hr] where we can:

$$\text{Frequency} \quad \left[\frac{1}{\cancel{hr}}\right] \times \left[\frac{\cancel{hr}}{yr}\right] = \left[\frac{1}{yr}\right]$$

You can see the equation is dimensionally correct.

Example

To convert a frequency of 5.00E-03 per hour, into a frequency per year, we must multiply by 8760 hrs/year:

Freq [1/yr]	=	5.00E-03 [1/hr] x	8760 [hr/yr]
	=	5.00E-03 [1/$\cancel{hr}$] x	8760 [$\cancel{hr}$/yr]
	=	5.00E-03 x 8760 [1/yr]	
	=	4.38E-01 [/yr]	

2. Dimensions

Example
To convert 1.3E-05 /hr to /yr:

$$1.3\text{E-}05\ [/\text{hr}] = 1.3\text{E-}05\ [/\text{hr}] \times 8760\ [\text{hr}/\text{yr}]$$

$$= 1.14\text{E-}01\ /\text{yr}$$

Example
To convert 7.11E-02 /yr to /hr:

$$7.11\text{E-}02\ [/\text{yr}] = \frac{7.11\text{E-}02\ [/\text{yr}]}{8760\ [\text{hr}/\text{yr}]}$$

$$= 8.12\text{E-}06\ /\text{hr}$$

In the following examples, assume that in one year, there are 8760 hours, 365 days and 12 months.

Example 5: express the following frequencies /year:

a)   5.00E-01 /hr
b)   2.0E-05 /hr
c)   800 /hr
d)   0.00001 /hr
e)   2 /day

Example 6: express the following frequencies /hour:

a)   5.00E-01 /yr
b)   2.0E-05 /yr
c)   3 / day
d)   0.0001 /yr
e)   8.76E-02 /yr

Intervals and Frequencies

The interval between two events, T is equal to the reciprocal of the event frequency, F.

$$T = 1/F$$

And similarly, the frequency of an event F, is equal to the reciprocal of the interval between the events, T.

$$F = 1/T$$

Say, if you take the dog for a walk twice per day, the frequency of the walk is 2 (/day). The interval between walks is ½ day.

If the frequency is /year, the interval is expressed in years, or if the frequency is expressed in /hour, the interval is in hours.

Example

A process upset occurs every 780 hours. What is the upset frequency?

The time between process upsets, T is 780 hrs.
The frequency of the upset F, is given by:

$$F = 1/T$$
$$= 1/780 \text{ hr}$$
$$= \frac{1}{780} \left[\frac{1}{hr} \right]$$
$$= 1.28\text{E-}03 \text{ /hr}$$

Example

An operator error occurs every 2 years. What is the error frequency?

$$\text{Freq} = 1/2 \text{ yr}$$
$$= 0.5 \text{ /yr}$$

If we wish to convert to /hr.

$$\text{Freq} = \frac{0.5}{8760} \left[\frac{1/yr}{hr/yr} \right]$$

$$= 5.71\text{E-}05 \text{ /hr}$$

2. Dimensions

Example 7: express the following as frequencies /year:

a) an event occurs every day
b) an event occurs every 3 months
c) an event occurs every 800 hrs
d) an event occurs every 10 yrs
e) an event occurs every 1000 yrs

Example 8: express the following as frequencies /hour:

a) an event occurs every day
b) an event occurs every 3 months
c) an event occurs every 800 hrs
d) an event occurs every 10 yrs
e) an event occurs every 1000 yrs

3. INITIATING EVENT FREQUENCIES

Initiating events

The frequencies of occurrence of the initiating events that can lead to hazardous scenarios are typically expressed in events per year. Dimensionally: [/yr].

The initiating events of hazardous scenarios may include control system failures, mechanical failures and process upsets and can either result from external events, e.g. environmental, equipment related failures (hardware or software), or human error.

Note that in these exercises, we will not consider external events as these can be complex and are often addressed in a separate risk assessment exercise.

For our calculations, we should try to list all potential initiating events. For example, loss of cooling could be an initiating cause of an overtemperature hazard but we may not have any experience of how often we could have loss of cooling and so determining how frequently it occurs might be difficult.

We might reason that loss of cooling may be caused by a number of different mechanisms. If we quantify the frequency of each of these individual contributing initiating causes this will enable us to quantify the frequency of loss of cooling. In this example, the frequency of loss of cooling is equal to the sum of the individual contributing frequencies.

Example

Loss of cooling can result from the following events, what is the frequency of loss of cooling?

coolant leak,	λ_L	=	0.05 /yr
pump failure,	λ_P	=	0.1 /yr
control failure,	λ_C	=	0.1 /yr
therefore loss of cooling, λ		=	0.25 /yr.

3. Initiating Event Frequencies

We can do a quick sensibility check on this frequency by calculating the interval between loss of cooling events to see whether it seems reasonable.

If the frequency λ = 0.25 /yr

The interval T = $1/\lambda$

= 1 / 0.25 [/yr]

= 4 yrs

This might be within our experience.

Enabling events

Enabling events consist of operations or conditions that do not themselves directly cause the hazardous scenario, but which must be present in order for the scenario to occur. Enabling events are expressed as probabilities, and can include modes of operation (e.g., startup or shutdown).

In such cases, an initiating event may be the combination of a failure which occurs when a certain condition exists, for example during start-up. So the initiating event frequency is a combination of the failure frequency AND the probability that it will occur during the enabling condition.

Example

An operator regularly doses a batch process and if he makes an error, incorrect dosing will cause a vessel over-pressure, potential rupture and explosion. Therefore, for the incorrect dosing to occur, the operator first has to carry out the dosing operation, and secondly, he has to make an error by dosing the incorrect amount.

The initiating event frequency for this scenario is therefore a function of how frequently the operator doses the batch (an enabling event frequency), and the chance that the operator will make an error (probability of error).

3. Initiating Event Frequencies

The initiating event frequency, λ_{EVENT} can be calculated from the dosing operation frequency, λ_{DOSE}, i.e. the number of batches run per year AND the probability that the dosing error is made during the batch operation, P_{ERROR}.

Therefore:

$$\lambda_{EVENT} = \lambda_{DOSE} \times P_{ERROR}$$

If the batch operation is run 4 times per year, $\lambda_{DOSE} = 4$ /yr and the probability of operator error P_{ERROR}, is 1%, then the hazardous event frequency λ_{EVENT}, can be calculated as:

$$\begin{aligned}
\lambda_{EVENT}\ [/yr] &= 4\ [/yr] \times 1\% \\
&= 4\ [/yr] \times 0.01 \\
&= 0.04\ [/yr] \quad \text{(or 4.00E-02 /yr)}
\end{aligned}$$

Note that if the number of batches per year changes, then the risk of vessel rupture also changes.

Example 9:

A maintainer operates some manual valves during annual shutdown. A vessel overfill will occur if the valves aren't returned to their correct position before the process is re-started. If the probability of maintainer error is 10%, what is the expected frequency of overfill?

Example 10:

An operator overrides a safety function to enable the start-up of a process after an annual shutdown. There is a 1 in 10 chance he will fail to remove the override after start-up, which will lead to an overtemperature hazard. What is the expected frequency of overtemperature?

Example 11:

An operator adds a catalyst to a reactor vessel at the start of a weekly batch run. Tests have shown that there is a 0.1% chance that the catalyst contains impurities that will cause an exothermic reaction. What is the expected frequency of this reaction?

3. Initiating Event Frequencies

Example 12:

Maintainers clean sludge from a tank after every batch run. The cleaning takes about 4 hours at the end of each day. If an inerting valve fails open, nitrogen will flow into the vessel and asphyxiate the maintainers. The inerting valve fail open failure rate is extimated to be 2.0E-06 /hr. What is the expected frequency of the hazard?

Example 13:

A batch process is run every day and the reaction is exothermic for the first two hours. If loss of cooling occurs during the exothermic phase, then a runaway reaction will result. Possible causes of loss of cooling are: coolant leak, estimated to occur every 4 years; and control failure, once every 10 years. What is the frequency of runaway reaction?

Example 14:

While moving gas cylinders to a nitrogen hookup station, an operator drops an uncapped cylinder, resulting in the valve breaking off and releasing gas. Cylinders are moved every 21 days and it is estimated that there is a 1% chance of dropping an uncapped cylinder. What is the frequency of gas release?

Example 15:

In example 14, what is the frequency of gas release if only 1 in 20 cylinders are uncapped? Note: fitting of caps will protect the cylinder valve.

Example 16:

A coastal tanker delivers fuel to a tank storage depot once per month. A dangerous undetected failure of an inventory control system, indicates that there is sufficient room in the tank and the fuel is unloaded but the tank overflows. The inventory control system is tested every year but estimated to fail approximately every 10 years. What is the frequency of tank overflow?

3. Initiating Event Frequencies

Example 17:

A fuel tank leaks from seals at a frequency of approximately 5.0E-02 /yr. If a leak occurs, the probability of ignition and subsequent explosion, is 3%. What is the frequency of explosion?

Example 18:

If the explosion in example 17 occurs, there will be operator casualties if the explosion occurs during working hours. The site is manned from 8am until 6pm every weekday. What is the expected casualty frequency?

Example 19:

Every Friday morning, kerosene is delivered from a road tanker to a storage depot. The delivery operation lasts 2 hours and hose failure during this time would result in a leak with the possibility of operator injury. Hose failures are expected to occur on average, every 10 years. What is the expected leak frequency?

Example 20:

There is a level control valve on both the import and export lines of a separator vessel. The control valves regulate the flow into and out of the vessel to maintain a working level. Each control valve has a fail-to-regulate failure rate of 6.0E-05 /hr. If the inlet valve sticks (fails to regulate), the failure will not be detected as level control can still be achieved by operation of the export valve. Both valves are fully tested every year.

    a) What frequency can we expect the valves to contribute to failure of level control?

    b) What is the expected failure frequency of the level control assuming the process control system failure rate is 0.1 /yr?

4. THE ALARP PRINCIPLE

Note: refer to section 5 of [11.1].

Value of Preventing a Statistical Fatality, VPF

It is possible that a safety improvement measure will not eliminate a risk, but reduce the risk by an amount and we will need to evaluate the risk reduction provided as the benefit against the cost of implementation.

Some organisations operate a cost per life saved target (or Value of Preventing a Statistical Fatality: VPF).

The cost of preventing fatalities over the life of the plant is compared with the target VPF. The improvements will be implemented unless costs are grossly disproportionate.

Example

Question: Application of ALARP

A £2M target VPF is used in a particular industry. A maximum tolerable risk target of 10^{-5} pa has been established for a particular hazard, which is likely to cause 2 fatalities.

The proposed safety system has been assessed and a risk of 8.0×10^{-6} pa predicted. Given that the Broadly Acceptable (Negligible) Risk is 10^{-6} pa then the application of ALARP is required.

In this example, for a cost of £10,000, additional instrumentation and redundancy will reduce the risk to 2.0×10^{-6} pa (just above the negligible region) over the life of the plant which is 30 yrs.

Should the proposal be adopted?

Answer: Number of lives saved over the life of the plant is given by:

$$N = \text{(reduction in the fatality frequency)} \times \text{number of fatalities per incident} \times \text{life of the plant}$$

$$= (8.0 \times 10^{-6} - 2.0 \times 10^{-6}) \times 2 \times 30$$

$$= 3.6 \times 10^{-4}$$

4. The ALARP Principle

Hence the cost per life saved is:

VPF $\quad=\quad$ £10000 / 3.6 x10^{-4}

$\quad\quad\quad=\quad$ £27.8M

The calculated VPF is >10 times the target cost per life saved criterion of £2M and therefore the proposal should be rejected.

Example

Question: Application of ALARP

A £4M target VPF is used and maximum tolerable risk target of 10^{-4} pa has been established for a particular hazard, which is likely to cause a single employee fatality.

An assessment of the proposed safety system has predicted a risk of 3.0x10^{-5} pa which meets the maximum tolerable risk target. However, for a cost of £100,000, additional redundancy will reduce the risk to 8.0 x10^{-6} pa over the life of the plant which is 25 yrs.

Should the proposal be adopted?

Answer: Number of lives saved over the life of the plant is given by:

N $\quad=\quad$ (reduction in the fatality frequency) x number of fatalities per incident x life of the plant

$\quad\quad=\quad$ (3.0 x10^{-5} - 8.0 x10^{-6}) x 1 x 25

$\quad\quad=\quad$ 2.2 x10^{-5} x 1 x 25

$\quad\quad=\quad$ 5.5 x10^{-4}

Hence the cost per life saved is:

VPF $\quad=\quad$ £100,000 / 5.5 x10^{-4}

$\quad\quad\quad=\quad$ £181.8M

The calculated VPF is >10 times the target cost per life saved criterion of £4M and therefore the proposal should be rejected.

4. The ALARP Principle

Example

Question: Application of ALARP

A £4M target VPF is used and maximum tolerable risk target of 10^{-4} pa has been established for a particular hazard, which is likely to cause a single employee fatality.

An assessment of the proposed safety system has predicted a risk of 9.0×10^{-5} pa which meets the maximum tolerable risk target. However, for a cost of £50,000, additional redundancy will reduce the risk to 4.0×10^{-6} pa over the life of the plant which is 25 yrs.

Should the proposal be adopted?

Answer: Number of lives saved over the life of the plant is given by:

$$N = \text{(reduction in the fatality frequency)} \times \text{number of fatalities per incident} \times \text{life of the plant}$$

$$= (9.0 \times 10^{-5} - 4.0 \times 10^{-6}) \times 1 \times 25$$

$$= 8.6 \times 10^{-5} \times 1 \times 25$$

$$= 2.15 \times 10^{-3}$$

Hence the cost per life saved is:

$$\text{VPF} = £50,000 / 2.15 \times 10^{-3}$$

$$= £23.3M$$

The calculated VPF is <10 times the target VPF criterion of £4M and therefore the proposal should be accepted.

4. The ALARP Principle

Example 21: should the following proposals be rejected or accepted:

a) The target VPF is £4M, the proposed safety system has predicted single fatality risk of 9.0×10^{-5} pa which meets the maximum tolerable risk target. However, for a cost of £50,000, additional redundancy will reduce the risk to 5.0×10^{-5} pa over the life of the plant which is 25 yrs.

b) The target VPF is £4M, the proposed safety system has predicted risk which results in 3 fatalities, of 8.0×10^{-6} pa which meets the maximum tolerable risk target. However, for a cost of £100,000, additional instrumentation and redundancy will reduce the risk to 3.0×10^{-6} pa over the life of the plant which is 30 yrs.

c) The target VPF is £2M, the proposed safety system has predicted single fatality risk of 5.5×10^{-5} pa which meets the maximum tolerable risk target. However, for a cost of £30,000, additional instrumentation will reduce the risk to 1.2×10^{-6} pa over the life of the plant which is 30 yrs.

d) The target VPF is £2M, the proposed safety system has predicted risk which results in 2 fatalities, of 8.0×10^{-6} pa which meets the maximum tolerable risk target. However, for a cost of £1,000, modified procedures will reduce the risk to 3.0×10^{-6} pa over the life of the plant which is 30 yrs.

e) The target VPF is £2M, the proposed safety system has predicted risk which results in 5 fatalities, of 3.0×10^{-6} pa which meets the maximum tolerable risk target. However, for a cost of £10,000, additional redundancy will reduce the risk to 1.1×10^{-6} pa over the life of the plant which is 25 yrs.

f) The target VPF is £2M, the proposed safety system has predicted single fatality risk of 2.0×10^{-5} pa which meets the maximum tolerable risk target. However, for a cost of £1,000, additional maintenance will reduce the risk to 7.0×10^{-6} pa over the life of the plant which is 30 yrs.

4. The ALARP Principle

g) The target VPF is £1M, the proposed safety system has predicted single fatality risk of 4.0×10^{-5} pa which meets the maximum tolerable risk target. However, for a cost of £250,000, additional redundancy will reduce the risk to 5.0×10^{-6} pa over the life of the plant which is 25 yrs.

h) The target VPF is £1M, the proposed safety system has predicted risk which results in 2 fatalities, of 9.0×10^{-6} pa which meets the maximum tolerable risk target. However, for a cost of £10,000, additional instrumentation will reduce the risk to 4.0×10^{-6} pa over the life of the plant which is 30 yrs.

i) The target VPF is £1M, the proposed safety system has predicted risk which results in 5 fatalities, of 8.76×10^{-6} pa which meets the maximum tolerable risk target. However, for a cost of £5,000, additional safety barriers will reduce the risk to 3.2×10^{-6} pa over the life of the plant which is 25 yrs.

j) The target VPF is £1M, the proposed safety system has predicted single fatality risk of 4.0×10^{-5} pa which meets the maximum tolerable risk target. However, for a cost of £100,000, additional redundancy will reduce the risk to 5.0×10^{-6} pa over the life of the plant which is 30 yrs.

5. DETERMINING SAFETY INTEGRITY LEVEL (SIL) TARGETS

Note: refer to section 6 of [11.1].

Risk Reduction Factor (RRF)

Example, if a smoke alarm works for 9 fires out of 10, then it has a Probability of Failure on Demand (PFD) of 0.1. A PFD of 0.1 would reduce the hazard frequency by a factor of 10, giving a Risk Reduction Factor (RRF) of 10.

$$\text{Summarising, PFD} = 1 / \text{RRF}.$$

It is useful to remember that mathematically, PFD is a probability and therefore has a value between zero and 1. It is also dimensionless, which means that simple dimensional checks can be carried out on any SIL target calculations we do.

Example 22: in the following examples, convert PFDs to RRFs and vice versa:

a) $PFD = 0.11$;

b) $PFD = 10^{-2}$;

c) $PFD = 3.33\ E\text{-}03$;

d) $PFD = 2 \times 10^{-2}$;

e) $RRF = 1000$;

f) $RRF = 4 \times 10^3$;

g) $RRF = 200$;

h) $RRF = 45$.

5. Determining SIL Targets

Example 23: in the following examples, what SIL can be claimed for the given PFD or RRF:

a) PFD = 0.11;

b) PFD = 10^{-2};

c) PFD = 3.33 E-03;

d) PFD = 2×10^{-2};

e) RRF = 1000;

f) RRF = 4×10^3;

g) RRF = 200;

h) RRF = 45.

Modes of Operation (Demand and Continuous Mode Systems)

In determining the way in which a Safety Instrumented Function (SIF) operates, IEC61511-1, 3.2.43 offers the following definitions.

Demand Mode: where a specified action is taken in response to process conditions or other demands. In the event of a dangerous failure of the SIF a potential hazard only occurs in the event of a failure of the process of BPCS.

Continuous Mode: where in the event of a dangerous failure of the SIF a potential hazard will occur without further failure unless action is taken to prevent it.

5. Determining SIL Targets

Demand Mode SIL Targets

IEC61511-1, 9.2.4 groups PFD targets into bands, or Safety Integrity levels (SILs). In the above example, we have a PFD target of $<10^{-1}$ for our safety function, and this gives a SIL1 requirement as shown, Table 1.

TABLE 1. DEMAND MODE SIL TARGETS

Demand Mode of Operation (Average probability of failure to perform its design function on demand)	Safety Integrity Level
$\geq 10^{-5}$ to $<10^{-4}$	4
$\geq 10^{-4}$ to $<10^{-3}$	3
$\geq 10^{-3}$ to $<10^{-2}$	2
$\geq 10^{-2}$ to $<10^{-1}$	1

Note: the PFD target is grouped into SIL bands because the standard requires an appropriate degree of rigour in the techniques and measures that are applied in the control and avoidance of systematic failures.

Demand Mode Safety Function Example

<u>Question</u>: A process area is manned for 2 hours per day. Overpressure of the process will result in a gas leak and it is estimated that 1 in 10 gas leaks will cause an explosion that will result in the death of the operator.

Analysis indicates that the overpressure condition will occur every 5 yrs (a rate of 0.2 pa). Assume that the maximum tolerable frequency for the hazard (operator killed by explosion) is 10^{-4} pa.

What is the required PFD of the SIF?

5. Determining SIL Targets

Answer:

The fatality rate is:

$$= 0.2 \text{ /yr} \times 2/24 \times 1/10$$

$$= 1.67 \times 10^{-3} \text{ /yr}$$

Therefore the safety system must have a PFD of:

$$= 10^{-4} \text{ /yr} / 1.67 \times 10^{-3} \text{ /yr}$$

$$= 6.0 \times 10^{-2}, \text{ which is equivalent to SIL1.}$$

This is an example of a demand mode SIF in that it is only called upon to operate at a frequency determined by the failure rate of the equipment under control.

We can confirm that the result is actually a PFD because we have divided a rate by a rate to give a dimensionless quantity, i.e. a probability.

Example 24: in the following examples, determine if a SIF is required, and if so, what is its PFD?

    a)   A process area is manned for 2 hours per day. Overtemperature of the process will result in an explosion that will result in the death of the operator. Analysis indicates that the overtemperature condition will occur every 20yrs.

    b)   A reactor vessel is single-manned 24hrs a day in three 8 hour shifts. An exothermic reaction as a result of the presence of impurities has been observed to occur about once per year. If left unchecked, the pressure increase would result in a vessel rupture and the death of the operator. The control system can provide protection on 90% of occasions by detecting the high temperature and quenching the reaction. A PSV, with a PFD of 0.1 provides pressure relief on the vessel.

5. Determining SIL Targets

c) A process vessel is manned 4hrs a day. A high liquid level as a result of a level control failure will overflow, ignite and result in a single fatality. The initiating event frequency is 0.1/yr and the probability of ignition has been estimated at 35%.

d) Upstream pipeline overpressure events are predicted to occur every 3 months and a pipeline rupture would result in the deaths of 5 maintenance staff. The pressure control system can compensate for upstream events 9 times in 10 and if an overpressure occurs, there is a 10% chance that it will rupture the pipe.

e) A maintainer opens and closes a valve every month. After he opens the valve, there is a 1% chance he will forget to close it and a tank will overfill. A high level alarm with a PFD of 0.1, is provided and there are periodic manual checks, which are estimated to prevent an overfill 9 times in 10. If the tank does overflow, the resulting vapour cloud will drift off site and find an ignition source. Multiple third party fatalities will result.

f) An export valve failing closed will result in liquid carry-over into a compressor resulting in £50M of damage and lost revenue. The valve failure rate is 2.00E-06 /hr and the maximum tolerable frequency for the event is 10^{-4} /yr. The DCS can provide some protection by closing the import valve on high level.

g) A process area is manned for 2 hours per day. A release of caustic material leading to the operator fatality could be caused by the following, with the corresponding failure rates:
    - Acid Storage Tank Rupture, λ = 3.0E-06 /yr;
    - Control System Failure, λ = 0.1 /yr;
    - Level Transmitter (Fails Low) , λ = 1.45E-02 /yr;
    - Pipe Failure, λ = 1.3E-05 /yr.

5. Determining SIL Targets

h) Hydrocarbon leak following rig loss of position due to either loss of anchors estimated at 1 in 100 yrs, or collision with another vessel, estimated at 3 in 100 yrs. The maximum tolerable frequency for the event is 10^{-4} /yr.

i) A chemical process is manned for 2 hours a day by a team of 2 skilled operators. An increase in temperature, as a result of an upstream process upset, will occur at a frequency of 3.60E-01 / year, and will result in a harmful vapour being produced which can kill all personnel in the vicinity if inhaled. Operators wear the necessary protective equipment, including a gas mask which is deemed to be 90% effective. A PLC based control system monitors temperature and can introduce an inhibitor if the temperature increases too much.

j) An operator manually fills a storage tank with kerosene every month. There is a 1% chance that he will not estimate the available capacity correctly and cause the tank to overfill. If the tank overflows, there is a 1% probability that the kerosene will ignite and if it does, the operator has a 50:50 chance of survival.

Continuous Mode SIL Targets

IEC61511-1, 9.2.4 also provides SIL Targets for continuous mode systems, Table 2.

TABLE 2. CONTINUOUS MODE SIL TARGETS

Continuous Mode of Operation (Probability of dangerous failure per hour, PFH)	Safety Integrity Level
$\geq 10^{-9}$ to $<10^{-8}$	4
$\geq 10^{-8}$ to $<10^{-7}$	3
$\geq 10^{-7}$ to $<10^{-6}$	2
$\geq 10^{-6}$ to $<10^{-5}$	1

Note: the target failure measure for continuous mode targets is a failure PFH or failure rate.

Example 25: in the following examples, what SIL can be claimed for the given PFH:

a) PFH = 0.11 x 10^{-6};

b) PFH = 10^{-6};

c) PFH = 3.33 E-07;

d) PFH = 2 x 10^{-6};

e) PFH = 10^{-5};

f) PFH = 4 x 10^{-6};

g) PFH = 200 x 10^{-6};

h) PFH = 45 x 10^{-7}.

Continuous Mode Safety Function Example

Figure 1 presents a simple example of a continuous mode safety function. The chemical in the boiler is heated by an electric element which is controlled by a temperature transmitter measuring the outlet.

FIGURE 1. CONTINUOUS MODE SAFETY FUNCTION

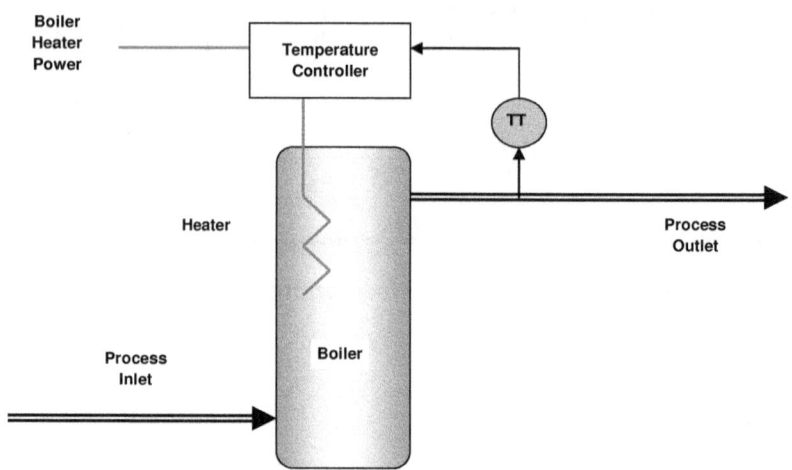

Assume that overheating of the boiler leads to rupture, chemical release and subsequent fire, with the potential for a fatality. There is clearly a risk that should be managed. In this example, the failure rate of the entire process should not exceed the maximum tolerable risk for the hazard.

Question: Assume failure of the boiler leads to overheating and fire and that 1 in 400 failures leads to a fatality. Assume also that the maximum tolerable fatality rate is 10^{-5} pa (third party fatality).

What is the maximum tolerable failure rate of the boiler?

Answer: Since 1 in 400 failures must be less than or equal to the maximum tolerable risk, we can say:

$$10^{-5} \text{ pa} \geq \lambda_B \times 1/400$$

Where λ_B is the failure rate of the boiler.

Therefore:

λ_B = 400 x 10^{-5} pa

= 4.0 x10^{-3} pa, which is equivalent to SIL2.

This is an example of a continuous mode SIF that is continually at risk, i.e. in continuous use. The boiler system is allowed to fail 400 times more frequently than the maximum tolerable failure rate because only 1 in 400 failures results in the fatality.

In this example, we would have to design and build the process, i.e. the boiler, heating element and temperature sensor to SIL2 and the failure rate would have to be less than 4.0 x10^{-3} pa.

Example 26: in the following examples, determine the required PFH of the EUC control system.

a) A process area is manned for 2 hours per day. Overtemperature caused by failure of the process control will result in an explosion that will result in the death of the operator.

b) A reactor vessel is single-manned 24hrs a day in three 8 hour shifts. An exothermic reaction as a result of the control system failure would result in a vessel rupture and the death of the operator. A PSV, with a PFD of 0.01 provides pressure relief on the vessel.

c) A process vessel is manned 4hrs a day. A high liquid level as a result of a level control failure will overflow, ignite and result in a single fatality. The probability of ignition has been estimated at 35%.

d) Failure of a pressure control system would result in a pipeline rupture and the possible deaths of 2 maintenance staff if they are in the area. The maintenance staff are in the area and at risk for only 10mins per day.

5. Determining SIL Targets

e) Failure of a boiler control system would result in a fire and explosion. It is estimated that 1 in 200 explosions will lead to the deaths of 5 maintenance.

f) Failure of a burner management system on start-up could lead to a shortened purge duration and the ignition of fuel gas in a burner. One in 10 failures would result in an explosion with asset damage which can be tolerated at 10^{-4} /year.

g) A maintainer opens and closes a valve during annual maintenance. After he opens the valve, there is a chance that he will forget to close it again and a tank will overfill resulting in a vapour cloud, and a 1% chance of ignition and single operator fatality. What is the maximum allowed probability that the maintainer will forget to close the valve?

6. RELIABILITY TECHNIQUES

Note: refer to section 11 of [11.1].

Introduction

This section provides a brief introduction to basic reliability techniques. It is by no means comprehensive nor is it in any way new or unconventional but the methods described herein are routinely used by reliability engineers.

Definitions

For convenience, an abridged list of key terms and definitions is provided. More comprehensive definitions of terms and nomenclature can be found in many standard texts on the subject.

Availability – A measure of the degree to which an item is in the operable and committable state at the start of the mission, when the mission is called for at an unknown state.

Failure – The event, or inoperable state, in which an item, or part of an item, does not, or would not, perform as previously specified.

Failure rate – The total number of failures within an item population, divided by the total number of life units expended by that population, during a particular measurement interval under stated conditions.

Mean time between failures (MTBF) – A basic measure of reliability for repairable items: the mean number of life units during which all parts of the item perform within their specified limits, during a particular measurement interval under stated conditions.

Reliability – The probability that an item can perform its intended function for a specified interval under stated conditions.

Failure rate and mean time between/to failure (MTBF/MTTF).

The purpose of quantitative reliability measurements is to define the rate of failure relative to time and to model that failure rate in a mathematical distribution. The most basic quantitative reliability measure is failure rate, which is estimated using the following equation:

$$\lambda = F/T$$

Where: λ = Failure rate (sometimes referred to as the hazard rate);

Note: λ is the Greek letter lambda, frequently used to denote failure rate.

T = total number of device hours (running time/cycles/miles/etc.) during an investigation period for both failed and non-failed items;

F = the total number of failures occurring during the investigation period.

For example, if five electric motors operate for a collective total time of 50 years with five functional failures during the period, the failure rate is 0.1 failures per year.

Another very basic concept is the mean time between/to failure (MTBF/MTTF). The only difference between MTBF and MTTF is that we employ MTBF when referring to items that are repaired when they fail. For items that are simply thrown away and replaced, we use the term MTTF. The computations are the same.

The basic calculation to estimate mean time between failure (MTBF) and mean time to failure (MTTF), is the reciprocal of the failure rate function. It is calculated using the following equation.

$$\theta = T/F$$

Where: θ = Mean time between/to failure;

Note: θ is the Greek letter theta, sometimes used to denote MTBF.

T = Total running time/cycles/miles/etc. during an investigation period for both failed and non-failed items;

F = the total number of failures occurring during the investigation period.

6. Reliability Techniques

The MTBF for our industrial electric motor example is 10 years, which is the reciprocal of the failure rate for the motors. Incidentally, we would estimate MTBF for electric motors that are rebuilt upon failure. For smaller motors that are considered disposable, we would state the MTTF.

The failure rate is a basic component of many more complex reliability calculations. Depending upon the mechanical/electrical design, operating context, environment and/or maintenance effectiveness, a machine's failure rate as a function of time may decline, remain constant, increase linearly or increase geometrically. However, for most reliability calculations, a constant failure rate is assumed.

Example 27: in the following examples, determine the MTBF of the items.

a) A valve has a failure rate of 2×10^{-6} /hour.

b) A control function fails on average once in 10 years.

c) A diesel engine has a failure rate of 4.5×10^{-4} /hour.

d) A PLC has a failure rate of 2.2×10^{-6} /hour.

e) A loading arm has a failure rate of 2×10^{-2} /year.

f) A boiler has a failure rate of 2.2×10^{-5} /hour.

g) An interface module has a failure rate of 7×10^{-6} /hour.

h) A control function has a failure rate of 0.1 /year.

6. Reliability Techniques

Example 28: in the following examples, determine the MTBF of the item.

a) Two hundred valves each operate for 1,000 hours. During this time, there are 2 failures.

b) Twelve failures are reported in a fleet of diesel trucks that have accumulated a total of 750,000 operating hours.

c) My system contains 10 PLCs each with a failure rate of 2.2 x10^{-6} /hour.

d) One interface module operates for 1,000 hours, six operate for 2,000 hours and twenty operate for 4,000 hours. A total of 5 failures are reported.

e) Two boilers operate for 5 years without any failures. During the next year, one of the boilers fails.

Example 29: in the following examples, determine the failure rate of the item.

a) A valve has an MTBF of 24,700 hours.

b) A control function has an MTBF of 3 years.

c) A diesel engine has an MTBF of 5 years.

d) A PLC has an MTBF of 120,000 hours.

e) A loading arm has an MTBF of 2450 hours.

f) A boiler has an MTBF of 70,700 hours.

g) An interface module has an MTBF of 91,700 hours.

h) A control function has an MTBF of 35,000 hours.

6. Reliability Techniques

Basic mathematical concepts in reliability engineering

Many mathematical concepts apply to reliability engineering, particularly from the areas of probability and statistics. For most purposes, and for the purposes of these exercises, we'll limit our discussion to the exponential distribution.

The Exponential Distribution

The exponential distribution models items with a constant failure rate, representing the flat section of the bathtub curve. Most industrial machines spend most of their lives in the constant failure rate, so it is widely applicable.

Below is the basic equation for estimating the reliability of a machine that follows the exponential distribution, where the failure rate is constant as a function of time.

$$R(t) = \exp\{-\lambda \cdot t\}$$

Where: $R(t)$ = Reliability estimate for a period of time, cycles, miles, etc. (t); λ = Failure rate (1/MTBF, or 1/MTTF) and t = the time at risk.

In our electric motor example, if you assume a constant failure rate the likelihood of running a motor for six years without a failure, or the projected reliability, is 55 percent. This is calculated as follows:

$$\begin{aligned} R(t) &= \exp\{-0.1 \times 6\} \\ &= \exp\{-0.6\} \\ &= 0.5488 \\ &\approx 55\% \end{aligned}$$

Estimating System Reliability

Once the reliability of components or machines has been established relative to the operating context and required mission time, plant engineers must assess the reliability of a system or process. Again, for the sake of brevity and simplicity, we'll discuss system reliability estimates for series, parallel and shared-load redundant (M out of N) systems (MooN systems).

6. Reliability Techniques

Series Systems

Before discussing series systems, we should discuss Reliability Block Diagrams (RBDs). RBDs simply map a process from start to finish. For a series system, all subsystems are required to be operating for the system to be operating, [Figure 2].

FIGURE 2. SERIES SYSTEM

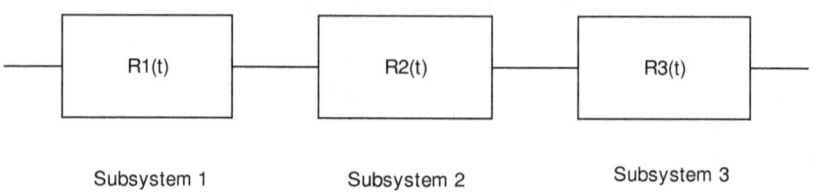

To calculate the system reliability for a serial process, you only need to multiply the estimated reliability of Subsystem 1 at time (t) by the estimated reliability of Subsystem 2 at time (t), and by the estimated reliability of Subsystem 3 at time (t). The basic equation for calculating the system reliability of a simple series system is:

Rs(t) = R1(t) . R2(t) . R3(t)

Where: Rs(t) – System reliability for given time (t); Rn(t) – Subsystem or sub-function reliability for given time (t).

So, for a simple system with three subsystems, or sub-functions, each having an estimated reliability of 0.90 (90%) at time (t), the system reliability is calculated as 0.90 X 0.90 X 0.90 = 0.729, or about 73%.

Example 30: in the following examples, what is the series system reliability:

  a) Two subsystems, each with a reliability of 0.95.
  b) Three subsystems, each with a reliability of 0.95.
  c) Three subsystems, two of which have a reliability of 99% and one with a reliability of 70%.
  d) Two subsystems, one with a reliability of 99% and one with a reliability of 70%.
  e) Three subsystems, one of which have a reliability of 99% and two with a reliability of 70%.

Parallel Systems

Often, design engineers will incorporate redundancy into critical machines. Reliability engineers call these parallel systems. The block diagram for a simple two component parallel system is shown in Figure 3.

FIGURE 3. PARALLEL SYSTEM

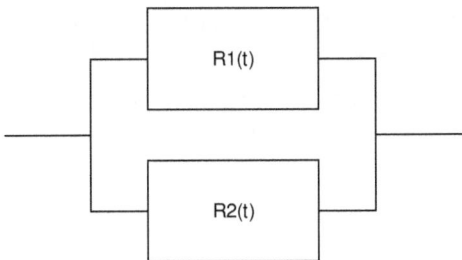

To calculate the reliability of redundant parallel system, where both machines are running, use the following simple equation:

Rs(t) = $1 - [\{1-R1(t)\} \cdot \{1-R2(t)\}]$

Where: Rs(t) – System reliability for given time (t); Rn(t) – Subsystem or sub-function reliability for given time (t).

For redundant systems, it may be easier to think in terms of unreliability. For this system, the system is unreliable if both subsystems are unreliable. Since unreliability U(t) = 1 – R(t) we can determine system unreliability as follows:

Us(t) = $U1(t) \cdot U2(t)$

Then substituting U(t) = 1 – R(t) we get:

1 - Rs(t) = $\{1-R1(t)\} \cdot \{1-R2(t)\}$

Therefore:

Rs(t) = $1 - [\{1-R1(t)\} \cdot \{1-R2(t)\}]$

The simple parallel system in our example with two components in parallel, each having a reliability of 0.90, has a total system reliability of 1 – (0.1 X 0.1) = 0.99. So, the system reliability is significantly improved from the component reliability.

6. Reliability Techniques

Example 31: in the following examples, what is the system reliability of a redundant system with the following components:

a) Two subsystems, each with a reliability of 0.95.
b) Three subsystems, each with a reliability of 0.95.
c) Three subsystems, two of which have a reliability of 99% and one with a reliability of 70%.
d) Two subsystems, one with a reliability of 99% and one with a reliability of 70%.
e) Three subsystems, one of which have a reliability of 99% and two with a reliability of 70%.

7. DANGEROUS AND SAFE FAILURES

Note: refer to section 11 of [11.1].

For reliability calculations to be meaningful, we are not only concerned with the failure rate of the system, but also how a system may fail, i.e. the failure mode.

Failure modes can be classified as safe or dangerous. Figure 4 shows a gas pipeline. If the pipeline is providing fuel to a power station and the Emergency Shutdown (ESD) Valve fails and spuriously closes, then the fuel supply is cut off and perhaps there will be some loss of revenue but the failure mode (fail closed) is a **safe** failure.

If the same valve fails in the open position, then we maintain the fuel supply but should there be an overpressure condition, we will not be able to isolate the fuel and make the pipeline safe. This failure mode (fail open) is therefore considered a **dangerous** failure.

FIGURE 4. EXAMPLE SAFETY INSTRUMENTED FUNCTION

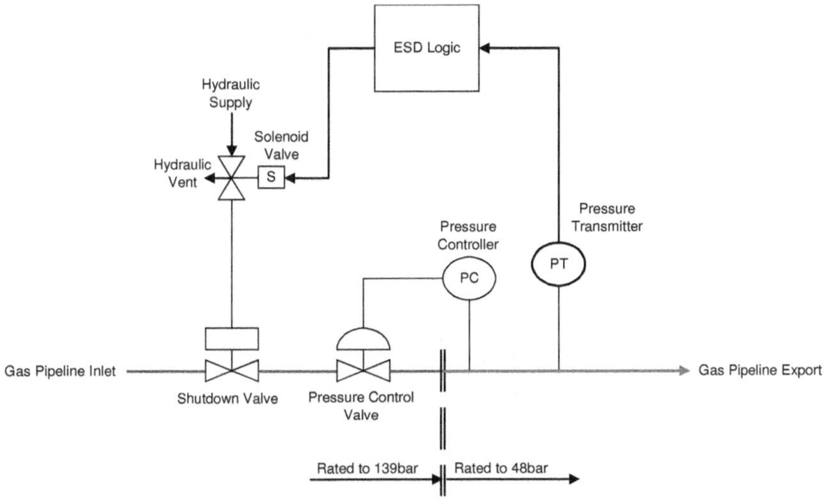

In this example, the fail open dangerous failure mode would not be revealed until a demand is placed upon it, i.e. until the valve is commanded to close. This is considered a dangerous undetected failure.

Alternatively, if the pipeline is providing a flow of coolant to the power station and the Shutdown Valve fails and spuriously closes, then the coolant is cut off and the power station may overheat. In this application, the same valve and the same failure mode (fail closed) is a **dangerous** failure. If the valve fails in the open position, then we maintain the coolant flow and therefore this failure mode (fail open) is therefore considered a **safe** failure.

A dangerous failure of a component in a safety instrumented function prevents that function from achieving a safe state when it is required to do so. **The dangerous failure rate is denoted by the symbol: λ_D.**

A safe failure does not have the potential to put the safety instrumented system in a dangerous or fail-to-function state but fails in such a way that it calls for the system to be shut down or the safety instrumented function to activate when there is no hazard present. **The safe failure rate is denoted by the symbol: λ_S.**

There may be failure modes which do not affect the safety function at all. These may include maintenance functions, indicators, data logging and other non-safety related (non-SR) functions. The non-SR failure rate is denoted by the symbol: λ_{non-SR}.

The total failure rate of an item, λ is equal to the sum of the safety-related and non-SR failure rates. Usually only λ_D and λ_S are included in reliability calculations.

$$\lambda = \lambda_D + \lambda_S + \lambda_{non-SR}$$

Detected and Undetected Failures

Dangerous failures that prevent the SIF from operating when required are either classified as detected failures, in that they are detected by diagnostics, or undetected failures that are not detected except by manual proof tests, which are typically performed annually.

So the dangerous failure rate λ_D can be apportioned between dangerous detected, λ_{DD} and dangerous undetected, λ_{DU}.

$$\lambda_D = \lambda_{DD} + \lambda_{DU}$$

7. Dangerous and Safe Failures

Question: A 4-20mA level transmitter, set to show a high reading on high level, has the following failure modes and failure rates.

Out of range signal >20mA	73×10^{-9} /hr
Loss of output, signal <4mA	5×10^{-9} /hr
Drift and read high	242×10^{-9} /hr
Drift and read low	75×10^{-9} /hr

If out of range signals are detected, what are the safe and dangerous detected and undetected failure rates if the safety function is to trip on high level?

Answer:

Out of range signal >20mA and Drift and read high:

Both out of range signal >20mA and drift and read high failure modes indicate a higher level than the actual level so therefore both of these failure modes are safe. The out of range signal is also detected but we can add both failure rates together to obtain the total safe failure rate and $\lambda_S = 73 \times 10^{-9}$ /hr $+ 242 \times 10^{-9}$ /hr $= 315 \times 10^{-9}$ /hr.

Drift and read low:

A low reading would indicate a lower level than the actual level so therefore this is a dangerous mode. If the signal is within the 4 to 20mA range there is no way of distinguishing a genuine low level from an incorrect low level. It is therefore dangerous undetected and $\lambda_{DU} = 75 \times 10^{-9}$ /hr.

Loss of output, signal <4mA:

A low reading also indicates a lower level than the actual level so this is also a dangerous mode. However, the failure would be detected because the signal is out of range, therefore dangerous detected and $\lambda_{DD} = 5 \times 10^{-9}$ /hr.

Summarising:

Out of range signal >20mA	73×10^{-9} /hr	Safe
Loss of output, signal <4mA	5×10^{-9} /hr	Dangerous Detected
Drift and read high	242×10^{-9} /hr	Safe
Drift and read low	75×10^{-9} /hr	Dangerous Undetected

7. Dangerous and Safe Failures

Example 32:

A 4-20mA level transmitter, configured to trip on low level and set to show a high current reading on high level, has the following failure modes and failure rates.

Out of range signal >20mA	73×10^{-9} /hr
Loss of output, signal <4mA	5×10^{-9} /hr
Drift and read high	242×10^{-9} /hr
Drift and read low	75×10^{-9} /hr

If out of range signals are detected, classify the data provided into safe, dangerous detected and dangerous undetected failure rates.

Example 33:

A 4-20mA pressure transmitter, configured to trip on high pressure and set to show a high current reading on high pressure, has the following failure modes and failure rates.

Out of range signal >20mA	17×10^{-9} /hr
Loss of output, signal <4mA	43×10^{-9} /hr
Drift and read high	48×10^{-9} /hr
Drift and read low	14×10^{-9} /hr

If out of range signals are detected, classify the data provided into safe, dangerous detected and dangerous undetected failure rates.

Example 34:

A pressure switch, configured to open on high pressure, has the following failure modes and failure rates.

Stuck closed	61×10^{-9} /hr
Spuriously opens	84×10^{-9} /hr

Classify the data provided into safe, dangerous detected and dangerous undetected failure rates.

7. Dangerous and Safe Failures

Example 35:

A slam-shut ESD Valve, configured to close on trip, has the following failure modes and failure rates.

Fail to close on demand	2.7×10^{-6} /hr
Fail to open on demand	1.3×10^{-6} /hr

Classify the data provided into safe, dangerous detected and dangerous undetected failure rates.

Example 36:

A 3-way Solenoid Valve, configured to vent when de-energised, has the following failure modes and failure rates.

Stuck closed	585×10^{-9} /hr
Spuriously opens	1010×10^{-9} /hr

Classify the data provided into safe, dangerous detected and dangerous undetected failure rates.

Example 37:

A safety relay, de-energise to trip, has the following failure modes and failure rates.

Coil open circuit	0.8×10^{-6} /hr
Contacts open circut	0.4×10^{-6} /hr
Contacts short circuit	0.6×10^{-6} /hr

Classify the data provided into safe, dangerous detected and dangerous undetected failure rates.

Example 38:

A gate valve, configured to close on trip, has the following failure modes and failure rates.

Fail to close on demand	3.3×10^{-6} /hr
Leak of theutility medium	2.9×10^{-6} /hr
Internal leakage	6.7×10^{-6} /hr

Classify the data provided into safe, dangerous detected and dangerous undetected failure rates.

7. Dangerous and Safe Failures

Example 39:

A 4-20mA temperature transmitter, configured to trip on high temperature and set to show a low current reading on high temperature, has the following failure modes and failure rates.

Out of range signal >20mA	77×10^{-9} /hr
Loss of output, signal <4mA	17×10^{-9} /hr
Drift and read high	2×10^{-9} /hr
Drift and read low	151×10^{-9} /hr

If out of range signals are detected, classify the data provided into safe, dangerous detected and dangerous undetected failure rates.

8. SYSTEM MODELLING

Note: refer to section 12 of [11.1].

Proof Test Period (T_p) and Mean Down Time (MDT)

If a failure occurs, it is assumed that on average it will occur at the mid-point of the test interval. In other words, the fault will remain undetected for 50% of the test period.

For both detected and undetected failures the Mean Down Time (MDT) depends upon the test interval and also the mean time to restoration, or MTTR.

The MDT is therefore calculated from:

$$\text{MDT} = \frac{\text{test interval}}{2} + \text{MTTR}$$

The MDT for detected failures therefore approximates to the repair time, since the test interval (autotest) is generally short compared to the MTTR. For undetected failures the repair time is short compared to the test interval, the Proof Test period Tp, and therefore MDT for undetected failures approximates to Tp/2.

Average Probability of Failure on Demand (PFD_{avg})

The simplified PFD_{avg} formulae for common configurations are presented in Table 3, for detected and undetected failures.

TABLE 3. PFD_{AVG} CALCULATION

Configuration	PFD_{DD}	PFD_{DU}
1oo1	$\lambda_{DD}.\text{MDT}$	$\lambda_{DU}.T_P/2$
1oo2	$\lambda_{DD}^2.\text{MDT}^2$	$\lambda_{DU}^2.T_P^2/3$
2oo2	$2.\lambda_{DD}.\text{MDT}$	$\lambda_{DU}.T_P$
1oo3	$\lambda_{DD}^3.\text{MDT}^3$	$\lambda_{DU}^3.T_P^3/4$
2oo3	$3.\lambda_{DD}^2.\text{MDT}^2$	$\lambda_{DU}^2.T_P^2$
3oo3	$3.\lambda_{DD}.\text{MDT}$	$3.\lambda_{DU}.T_P/2$
1oo4	$\lambda_{DD}^4.\text{MDT}^4$	$\lambda_{DU}^4.T_P^4/5$
2oo4	$4.\lambda_{DD}^3.\text{MDT}^3$	$\lambda_{DU}^3.T_P^3$

8. System Modelling

The total PFD$_{avg}$ is the sum of the detected and undetected components, i.e.

$$PFD_{avg} = PFD_{DD} + PFD_{DU}$$

Question: A 4-20mA level transmitter, has the following failure rates.

Dangerous Detected	5 x10^{-9} /hr
Dangerous Undetected	75 x10^{-9} /hr
Safe	315 x10^{-9} /hr

Calculate PFD$_{DD}$ and PFD$_{DU}$ for a simplex (1oo1) configuration, assuming MDT = 24 hrs and Tp = 1 year.

Answer:

PFD$_{DD}$ = λ_{DD}.MDT

= 5 x10^{-9} /hr x 24 hrs

= 1.2 x10^{-7}

PFD$_{DU}$ = λ_{DU}.T$_P$/2

= 75 x10^{-9} /hr x 8760 hrs

= 6.57 x10^{-4}

Example 40:

A slam-shut ESD Valve has the following failure rates.

λ_{DU}	61 x10^{-9} /hr
λ_S	84 x10^{-9} /hr

Calculate PFD$_{DD}$ and PFD$_{DU}$ for a simplex (1oo1) configuration, assuming MDT = 24 hrs and Tp = 1 year.

8. System Modelling

Example 41:

A pressure transmitter has the following failure rates.

λ_{DU} 16×10^{-9} /hr

λ_{DD} 81×10^{-9} /hr

λ_S 64×10^{-9} /hr

Calculate PFD_{DD} and PFD_{DU} for a simplex (1oo1) configuration, assuming MDT = 24 hrs and Tp = 1 year.

Example 42:

Calculate PFD_{avg} for example 41.

Example 43:

Using the data in example 41, calculate PFD_{DD} and PFD_{DU} for a 2oo3 configuration, assuming MDT = 24 hrs and Tp = 1 year.

Example 44:

Calculate PFD_{avg} for example 43.

Accounting for Common Cause Failures

Common cause failures (CCF) are failures that may result from a single cause but simultaneously affect more than one channel. They may result from a systematic fault for example, a design specification error or an external stress such as an excessive temperature that could lead to component failure in both redundant channels. It is the responsibility of the system designer to take steps to minimise the likelihood of common cause failures by using appropriate design practices.

The contribution of CCF in parallel redundant paths is accounted for by inclusion of a β-factor. The CCF failure rate that is included in the calculation is equal to β x the total failure rate of one of the redundant paths.

8. System Modelling

The following provides a summary of the basic forms.

For detected failures:
$PFD_{1oo1} = \lambda_{DD} \cdot MDT$
$PFD_{1oo2} = \lambda_{DD}^2 \cdot MDT^2 + \beta \cdot \lambda_{DD} \cdot MDT$

For undetected failures:
$PFD_{1oo1} = \lambda_{DU} \cdot T_p / 2$
$PFD_{1oo2} = \lambda_{DU}^2 \cdot T_p^2 / 3 + \beta \cdot \lambda_{DU} \cdot T_p / 2$

Where β is the contribution from common cause failures. T_p is the proof test interval and MDT is the Mean Down Time.

<u>Question</u>: A 4-20mA level transmitter, has the following failure rates.

Dangerous Detected	6×10^{-9} /hr
Dangerous Undetected	7×10^{-8} /hr
Safe	310×10^{-9} /hr

Calculate PFD_{avg} for a 1oo2 configuration, assuming MDT = 24 hrs, Tp = 1 year and β = 5%.

<u>Answer</u>:

PFD_{DD} = $\lambda_{DD}^2 \cdot MDT^2 + \beta \cdot \lambda_{DD} \cdot MDT$
= $(6 \times 10^{-9})^2 \times 24^2 + 0.05 \times 6 \times 10^{-9} \times 24$
= 7.2×10^{-9}

PFD_{DU} = $\lambda_{DU}^2 \cdot T_p^2 / 3 + \beta \cdot \lambda_{DU} \cdot T_p / 2$
= $(7 \times 10^{-8})^2 \times 8760^2 / 3 + 0.05 \times 7 \times 10^{-8} \times 8760 / 2$
= 1.55×10^{-5}

PFD_{avg} = $PFD_{DD} + PFD_{DU}$
= 1.55×10^{-5}

8. System Modelling

Example 45:

A pressure transmitter has the following failure rates.

λ_{DU} 16×10^{-9} /hr

λ_{DD} 81×10^{-9} /hr

λ_{S} 64×10^{-9} /hr

Calculate PFD_{avg} for a 2oo3 configuration, assuming MDT = 24 hrs, Tp = 1 year and β = 5%.

Example 46:

A shutdown valve has the following failure rates.

λ_{DU} 1.6×10^{-6} /hr

λ_{DD} 0.00 /hr

λ_{S} 6.4×10^{-6} /hr

Calculate PFD_{avg} for a 1oo3 configuration, assuming MDT = 24 hrs, Tp = 6 months and β = 5%.

9. REQUIREMENTS FOR HARDWARE FAULT TOLERANCE

Note: refer to section 12 of [11.1].

Approach

To address the requirements for Hardware Fault Tolerance (HFT), a quantitative assessment against Safe Failure Fraction (SFF) and Architectural Constraints is required.

Safe Failure Fraction

In the context of hardware safety integrity, the highest SIL that can be claimed for a safety function is limited by the HFT and the SFF, of the sub-systems that carry out that safety function.

A hardware fault tolerance of 1 indicates that the architecture of the sub-system is such that a dangerous failure of one of the sub-systems does not prevent the safety action from occurring i.e. a configuration of 1oo2 or 2oo3 would have a HFT of 1 or a configuration of 1oo3 or 2oo4 would have a HFT of 2.

Example 47: in the following configurations, what is the HFT?

a) 1oo1;

b) 1oo2;

c) 2oo3;

d) 1oo3;

e) 2oo4.

The following general relationships are used.

$$SFF = \Sigma (\Sigma \lambda_S + \Sigma \lambda_{DD}) / (\Sigma \lambda_S + \Sigma \lambda_D)$$

where:

$$\lambda_D = \lambda_{DU} + \lambda_{DD}$$

For each element in the safety function, SFF should be calculated. The value should then be used in Table 4 to determine SIL compliance for the level of hardware fault tolerance.

9. Requirements for Hardware Fault Tolerance

Example 48:

A slam-shut ESD Valve has the following failure rates.

λ_{DU} 61×10^{-9} /hr

λ_S 84×10^{-9} /hr

Calculate SFF.

Example 49:

A pressure transmitter has the following failure rates.

λ_{DU} 16×10^{-9} /hr

λ_{DD} 81×10^{-9} /hr

λ_S 64×10^{-9} /hr

Calculate SFF.

Example 50:

A shutdown valve has the following failure rates.

λ_{DU} 1.6×10^{-6} /hr

λ_{DD} 0.00 /hr

λ_S 6.4×10^{-6} /hr

Calculate SFF.

Architectural Constraints

IEC61511-1, 11.4.5 allows for assessment of hardware fault tolerance using the requirements of IEC61508-2, Tables 2 and 3.

Within IEC61508 [11], subsystems are categorised as either Type A or Type B. Generally, if the failure modes are well defined and the behaviour under fault conditions can be completely determined and there is sufficient adequate field data then the subsystem may be considered Type A. If any of these conditions may not be true then the subsystem must be considered Type B.

Simple mechanical devices such as valves are generally considered to be Type A. Logic Solvers are usually Type B in that they contain some processing capability and as such, their behaviour under fault conditions may not be completely determined. Sensors can be either

9. Requirements for Hardware Fault Tolerance

Type A or Type B depending on the technology and complexity of the device.

Example 51: in the following examples, should the components be considered Type A or Type B?

a) ESD Valve;
b) Pressure transmitter with a 4-20mA interface and internal diagnostics;
c) CPU Module;
d) Ethernet module;
e) Safety relay;
f) Solenoid operated valve;
g) Temperature switch;
h) Float valve.

9. Requirements for Hardware Fault Tolerance

The architectural constraints for a safety function are summarised in Table 4.

TABLE 4. ARCHITECTURAL CONSTRAINTS

Type A subsystems definition: Failure modes of all constituent parts well defined, and Behaviour of the subsystem under fault conditions completely determined, and Sufficient dependable data from field experience to show that the claimed failure rates for detected and undetected dangerous failures are met			
Safe Failure Fraction	Hardware Fault Tolerance (N)		
	0	1	2
<60%	SIL 1	SIL 2	SIL 3
60% - <90%	SIL 2	SIL 3	SIL 4
90% - <99%	SIL 3	SIL 4	SIL 4
≥ 99%	SIL 3	SIL 4	SIL 4
Type B subsystems definition: Failure mode of at least one constituent component is not well defined, or The behaviour of the subsystem under fault conditions cannot be completely determined, or There is insufficient dependable data from field experience to support the claimed failure rates for detected and undetected dangerous failures			
Safe Failure Fraction	Hardware Fault Tolerance (N)		
	0	1	2
<60%	Not allowed	SIL 1	SIL 2
60% - <90%	SIL 1	SIL 2	SIL 3
90% - <99%	SIL 2	SIL 3	SIL 4
≥ 99%	SIL 3	SIL 4	SIL 4

Note: A hardware fault tolerance of N means that N+1 faults could cause the loss of the safety function.

Example

In this example, Figure 5 the safety function consists of two level switches operating in a 1oo2 configuration. If either switch detects a high level, then the PLC will de-energise the Solenoid Operated Valve (SOV) which will allow the ESD Valve to close.

The assessment of the architectural performance requires that we first identify what type, i.e. A or B, each element is. This can generally be determined using the definitions provided in Table 4. As a general rule, you have to be sure of the failure modes and failure behaviour of an item, and have very good failure data in order to consider it a Type A. Otherwise; the item must be considered Type B.

Our failure data for each element, then allows us to calculate the SFF. The item type and SFF are tabulated below each element in Figure 5.

9. Requirements for Hardware Fault Tolerance

The redundancy refers to the level of fault tolerance for each element. The level switches operating in a 1oo2 configuration have a fault tolerance, or redundancy of 1. All other items have no fault tolerance and therefore have a redundancy of 0.

Finally, the SIL that can be claimed for the architectural performance of each element can be determined using this information in Table 4.

FIGURE 5. EXAMPLE SAFETY INSTRUMENTED FUNCTION

Type	A	B	A	A
SFF	0.40	0.95	0.72	0.25
HFT	1	0	0	0
Architectural SIL	2	2	2	1
Allowed SIL (Arch)	1			
Overall Allowed SIL	SIL1			

The level transmitters are type A, therefore the type A criteria applies. With a SFF of 0.74 and a fault tolerance of 1, the level transmitters comply with the architectural constraints of SIL3.

Similarly, the SOV and ESD Valve can also be assessed. The SOV, also type A, has a fault tolerance of 0 and a SFF of 0.72 giving SIL2. The ESD Valve, type A, fault tolerance of 0 and a SFF of 0.25 giving SIL1.

FIGURE 6. LEVEL SWITCH ARCHITECTURAL CONSTRAINTS

Type A subsystems definition: Failure modes of all constituent parts well defined, and Behaviour under fault conditions completely determined, and Sufficient dependable data from field experience to show that the claimed failure rates for detected and undetected dangerous failures are met.			
Safe Failure Fraction (SFF)	Hardware Fault Tolerance (N)		
	0	1	2
<60%	SIL1	SIL2	SIL3
60% - <90%	SIL2	SIL3	SIL4
90% - <99%	SIL3	SIL4	SIL4
≥99%	SIL3	SIL4	SIL4

The PLC has been considered a Type B device. This is true for most PLCs because as they are software controlled, there is an element of uncertainty about their failure behaviour and consequently, all of the

9. Requirements for Hardware Fault Tolerance

conditions required for Type A are not met. The assessment of the PLC must therefore be against the Type B requirements.

FIGURE 7. PLC ARCHITECTURAL CONSTRAINTS

Type B subsystems definition:
Failure modes of at least one constituent part not well defined, or
Behaviour of the subsystem under fault conditions cannot be completely determined, or
There is insufficient dependable data from field experience to support the claimed failure rates for detected and undetected dangerous failures.

Safe Failure Fraction (SFF)	Hardware Fault Tolerance (N)		
	0	1	0
<60%	Not allowed	SIL1	SIL2
60% - <90%	SIL1	SIL2	SIL3
90% - <99%	SIL2 ← PLC	SIL3	SIL4
≥99%	SIL3	SIL4	SIL4

To summarise, the architectural SIL performance for each element is shown in Figure 5 and the SIL that can be claimed for the whole safety function, is SIL1. The architectural SIL performance for the whole safety function is limited by the lowest SIL claimed.

Example 52: in the following examples, what SIL can be claimed for the components?

 a) ESD Valve with a SFF of 62% in a 1oo2 configuration;

 b) Pressure transmitter with a 4-20mA interface and internal diagnostics, with a SFF of 94% in a 1oo1 configuration;

 c) CPU Module with a SFF of 98%, in a 1oo1 configuration;

 d) Ethernet module with a SFF of 95%, in a 1oo2 configuration;

 e) Relay with a SFF of 75%, in a 1oo1 configuration;

 f) Solenoid operated valve with a SFF of 75%, in a 1oo2 configuration;

 g) Temperature switch with a SFF of 65%, in a 1oo2 configuration;

 h) Float valve with a SFF of 40%, in a 1oo1 configuration.

10. ANSWERS

Example 1, p.7

a) 1.20×10^2; b) 1.04×10^4; c) 5.00×10^{-4}; d) 1.00×10^0; e) 6.05×10^{-8}; f) 1.00×10^7; g) 5.30×10^7; h) 3.00×10^{-5}; i) 2.00×10^0; j) 1.2×10^{-7}; k) 1.00×10^{-3}.

Example 2, p.7

a) 0.3; b) 0.00002; c) 0.9; d) 0.0001; e) 300,000; f) 50,000; g) 3.0; h) 0.00000005; i) 0.000001; j) 3; k) 3.

Example 3, p.8

a) 4.00E+01; b) 5.40E+04; c) 3.00E-04; d) 1.00E+00; e) 6.00E-06; f) 1.01E+08; g) 3.50E+08; h) 1.30E-05; i) 2.00E+00; j) 1.00E-07; k) 1.00E-03.

Example 4, p.8

a) 0.5; b) 0.00002; c) 900; d) 0.0001; e) 300,000; f) 0.00005; g) 5.0; h) 0.00000004; i) 0.000001; j) 1.0; k) 3.0.

Example 5, p.10

a) 4380; b) 1.752×10^{-1}; c) 7.0×10^6; d) 8.76×10^{-2}; e) 730.

Example 6, p.10

a) 5.71×10^{-5}; b) 2.28×10^{-9}; c) 0.125; d) 1.14×10^{-8}; e) 10^{-5}.

Example 7, p.12

a) 365; b) 4; c) 10.95; d) 0.1; e) 10^{-3}.

10. Answers

Example 8, p.12

a) 4.17×10^{-2}; b) 4.57×10^{-4}; c) 1.25×10^{-3}; d) 1.14×10^{-5}; e) 1.14×10^{-7}.

Example 9, p.15

Frequency of maintenance action (opportunity), F_{maint} = 1 /yr.

Probability of error, P_{error} = 10%.

Frequency of overfill, $F_{overfill} = F_{maint} \times P_{error}$ = 1 /yr x 0.1 = 0.1 /yr.

Example 10, p.15

Frequency of override action (opportunity), $F_{override}$ = 1 /yr.

Probability of error, P_{error} = 0.1.

Frequency of hazard, $F_{hazard} = F_{override} \times P_{error}$ = 1 /yr x 0.1 = 0.1 /yr.

Example 11, p.15

Frequency of operator action (opportunity), F_{action} = 1 /week.

Probability of error, P_{error} = 0.1%

Frequency of reaction, $F_{reaction}$ = $F_{action} \times P_{error}$ = 1 /wk x 0.001

= 0.001 /wk

= 1.68×10^{-1} /hr.

Example 12, p.16

Frequency of valve failure, F_{valve} = 2.0×10^{-6} /hr.

Maintainers at risk for 4 hours in 24 hours, therefore the probability the tank will be occupied at any time, $P_{occupied}$ = 4/24 = 0.167.

Frequency of hazard, F_{hazard} = $F_{valve} \times P_{occupied}$

= 2.0×10^{-6} /hr x 0.167

= 3.33×10^{-7} /hr

10. Answers

Example 13, p.16

Reaction is exothermic for 2 hours in 24 hours, therefore the probability the reaction is at risk at any time, $P_{exothermic}$ = 2/24 = 0.083.

Frequency of coolant leak, F_{leak} = 0.25 /yr.

Frequency of control failure, $F_{control}$ = 0.1 /yr.

Therefore, frequency of loss of cooling, $F_{cooling}$ = $F_{leak} + F_{control}$

$$= 0.25 \text{ /yr} + 0.1 \text{ /yr}$$

$$= 0.35 \text{ /yr}$$

Frequency of reaction, $F_{reaction}$ = $F_{cooling} \times P_{exothermic}$

$$= 0.35 \text{ /yr} \times 0.083$$

$$= 2.92 \times 10^{-2} \text{ /yr}$$

Example 14, p.16

Interval between cylinder movements, $T_{movement}$ = 21 days

∴ Frequency of cylinder movement, $F_{movement}$ = 1 / 21 per day

$$= 0.0476 \text{ /day}$$

Probability of dropping cylinder, $P_{dropping}$ = 0.01

Frequency of valve breaking off leading to a gas release, F_{valve}

$$= F_{movement} \times P_{dropping}$$

$$= 0.0476 \text{ /day} \times 0.01$$

$$= 4.76 \times 10^{-4} \text{ /day}$$

Or, converting to per year F_{valve} = 4.76×10^{-4} /day x 365 days/year

$$= 1.74 \times 10^{-1} \text{ /year}$$

10. Answers

Example 15, p.16

From Ex.14, the frequency of valve breaking, $F_{valve} = 1.74 \times 10^{-1}$ /yr

Now, only 1 in 20 valve breakages lead to a release

∴ Frequency of gas release, $F_{release}$ = 1.74×10^{-1} /yr × 1 / 20

= 8.69×10^{-3} /yr

Example 16, p.16

This is a tricky question. If the inventory control system fails, we can assume it will remain in the failed condition until the failure is revealed either by testing or by a tank overflow.

In practice, this will be more likely by an overflow because the only failure that will be revealed by test is when the failure occurs in the month prior to the annual test being carried out.

Therefore, the frequency of overflow is approximately the frequency of control failure $F_{control} \approx 0.1$ /year. This answer would be acceptable.

More exactly, we should subtract from this the frequency of the failure occurring in the last month before test.

$F_{control}$ = 0.1 /yr − (0.1 /yr × 1/12)

= 0.1 − 0.1 × 0.083

= 0.1 − 0.0083

= 9.17×10^{-2} /yr

Which is very slightly less than 0.1 /yr.

10. Answers

Example 17, p.17

Frequency of seal leak, F_{leak} = 5.0E-02 /yr

Probability of ignition, $P_{ignition}$ = 0.03

Frequency of explosion, $F_{explosion} = F_{leak} \times P_{ignition}$

$$= 5.0E\text{-}02 \text{ /yr} \times 0.03$$

$$= 1.5E\text{-}03 \text{ /yr}$$

Example 18, p.17

From Ex.17, frequency of explosion, $F_{explosion}$ = 1.5E-03 /yr

Probability of occupancy, $P_{occupied}$ = 10 hrs/day x 5 days/week

$$= (10\text{hrs} / 24\text{hrs}) \times (5\text{days} / 7\text{days})$$

$$= 10 / 24 \times 5 / 7$$

$$= 0.416 \times 0.714$$

$$= 0.297$$

Frequency of casualty, $F_{casualty}$ = $F_{explosion} \times P_{occupied}$

$$= 1.5E\text{-}03 \text{ /yr} \times 0.297$$

$$= 4.5E\text{-}04 \text{ /yr}$$

10. Answers

Example 19, p.17

Frequency of hose failure, $F_{failure}$ = 0.1 /yr

Probability of fuel delivery, $P_{delivery}$ = 2 hrs/week

= 2 hrs / 168 hrs

= 0.012

Frequency of leak, F_{leak} = $F_{failure}$ × $P_{delivery}$

= 0.1 /yr × 0.012

= 1.2 × 10^{-3} /yr

Example 20, p.17

Failure of level control due to valve failure requires either the import valve to fail and be in the failed condition when the export valve fails, or the export valve fails and be in the failed condition when the import valve fails. As each valve is tested once per year, if a valve fails, it will remain in the failed state for on average half of the test interval, 6 months.

Therefore the failure rate of level control due to valve failure, F_{valves}

= F_{import} × (F_{export} × 6 months) + F_{export} × (F_{import} × 6 months)

As the failure rates are expressed in hours, we must convert the test interval, 6 months into 4380 hours.

Part a)

Then the frequency we expect the valves to contribute to failure of level control is given by:

F_{valves} = 6.0E-05 /hr × (6.0E-05/hr × 4380hr) + 6.0E-05 /hr × (6.0E-05/hr × 4380hr)

= 6.0E-05 /hr × 2.63E-01 + 6.0E-05 /hr × 2.63E-01

= 1.58E-05 /hr + 1.58E-05 /hr

= 3.15E-05 /hr

10. Answers

Part b)

Failure of the level control system, $F_{control}$ can be due to either failure of the process control, $F_{process}$ or failure of the valves, F_{valves}.

$F_{control}$ = $F_{process}$ + F_{valves}

= 0.1 /yr + 3.15E-05 /hr

Converting the process control failure rate into per hour:

$F_{control}$ = 1.14E-05 /hr = 3.15E-05 /hr

= 4.3E-05 /hr

Example 21, p.21

a) Number of lives saves in the life of the plant:

N = $(9.0 \times 10^{-5}$ /yr - 5.0×10^{-5}/yr) x 25yr

= 4.0×10^{-5} /yr x 25 yr

= 1.0×10^{-3}

Cost = £5 x10^4

Cost per life saved = £ 5 x10^4 / 1.0×10^{-3}

= £ 5.0 x10^7

= £50M

This is more than 10 times the VPF of £4M therefore the proposal should be rejected.

10. Answers

b) Number of lives saves in the life of the plant:

N = $3 \times (8.0 \times 10^{-6} / \text{yr} - 3.0 \times 10^{-6}/\text{yr}) \times 30\text{yr}$

= 4.5×10^{-4}

Cost = £100×10^{3}

Cost per life saved = £$1 \times 10^{5} / 4.5 \times 10^{-4}$

= £2.2×10^{8}

= £220M

This is more than 10 times the VPF of £4M therefore the proposal should be rejected.

c) Number of lives saves in the life of the plant:

N = 1.61×10^{-3}

Cost = £30×10^{3}

Cost per life saved = £18.6M therefore Accept.

d) Number of lives saves in the life of the plant:

N = 3.0×10^{-4}

Cost = £1×10^{3}

Cost per life saved = £3.3M therefore Accept.

e) Number of lives saves in the life of the plant:

N = 2.38×10^{-4}

Cost = £10×10^{3}

Cost per life saved = £42.1M therefore Reject.

10. Answers

f) Number of lives saves in the life of the plant:

N = 3.9 x10^{-4}

Cost = £1 x10^3

Cost per life saved = £2.5M therefore Accept.

g) Number of lives saves in the life of the plant:

N = 8.75 x10^{-4}

Cost = £250 x10^3

Cost per life saved = £286M therefore Reject.

h) Number of lives saves in the life of the plant:

N = 3.0 x10^{-4}

Cost = £10 x10^3

Cost per life saved = £33.3M therefore Reject.

i) Number of lives saves in the life of the plant:

N = 6.95 x10^{-4}

Cost = £5 x10^3

Cost per life saved = £7.2M therefore Accept.

j) Number of lives saves in the life of the plant:

N = 1.05 x10^{-4}

Cost = £100 x10^3

Cost per life saved = £95.2M therefore Reject.

10. Answers

Example 22, p.23

a) 9.09; b) 100; c) 300; d) 50; e) 1.00E-03; f) 2.50E-04; g) 5.00E-03; h) 2.22E-02.

Example 23, p.24

a) None; b) SIL1; c) SIL2; d) SIL1; e) SIL2; f) SIL3; g) SIL2; h) SIL1.

Example 24, p.24

Part a)

$F_{fatality}$ = $F_{overpressure}$ x P_{manned}

= 0.05 /yr x (2 hr / 24 hr)

= 0.05 /yr x 0.083

= 4.17E-03 /yr

which is greater than the MTR for a single fatality, 1.00E-04 /yr.

Therefore a SIF is required.

PFD = 1.00E-04 /yr / 4.17E-03 /yr

= 2.40E-02

Part b)

$F_{fatality}$ = $F_{impurity}$ x P_{manned} x $P_{control}$ x P_{PSV}

= 1 /yr x (8 hr / 24 hr) x 0.1 x 0.1

= 3.33E-04 /yr

which is greater than the MTR for a single fatality, 1.00E-04 /yr.

Therefore a SIF is required.

PFD = 3.00E-02

10. Answers

Part c)

$F_{fatality}$ = $F_{level} \times P_{manned} \times P_{ignition}$

= 0.1 /yr x (4 hr / 24 hr) x 0.35

= 5.83E-03 /yr

which is greater than the MTR for a single fatality, 1.00E-04 /yr.

Therefore a SIF is required.

PFD = 1.71E-02

Part d)

$F_{fatality}$ = $F_{overpressure} \times P_{control} \times P_{rupture}$

= 4 /yr x 0.1 x 0.1

= 4.00E-02 /yr

which is greater than the MTR for multiple fatalities, 1.00E-05 /yr.

Therefore a SIF is required.

PFD = 2.50E-04

Part e)

$F_{fatality}$ = $F_{valve\ operations} \times P_{error} \times P_{alarm} \times P_{checks}$

= 12 /yr x 0.01 x 0.1 x 0.1

= 1.20E-03 /yr

which exceeds the MTR for multiple third party fatalities, 1.00E-06 /yr.

Therefore a SIF is required.

PFD = 8.33E-04

Part f)

F_{damage} = $F_{valve} \times P_{DCS}$

= 1.75E-02 /yr x 0.1

= 1.75E-03 /yr

which is greater than the MTR for £50M damage, 1.00E-04 /yr.

Therefore a SIF is required.

PFD = 5.71E-02

Part g)

$F_{release}$ = $F_{rupture} + F_{control} + F_{transmitter} + F_{pipe}$

= (3.0E-06 + 0.1 + 1.45E-02 + 1.3E-05) /yr

= 1.15E-01 /yr

$F_{fatality}$ = $F_{release} \times P_{occupied}$

= 1.15E-01 /yr x 0.0833

= 9.54E-03 /yr

which is greater than the MTR for a single fatality, 1.00E-04 /yr.

Therefore a SIF is required.

PFD = 1.05E-02

10. Answers

Part h)

$$F_{\text{loss of position}} = F_{\text{anchors}} + F_{\text{collision}}$$
$$= 0.01 \text{ /yr} + 0.03 \text{ /yr}$$
$$= 0.04 \text{ /yr}$$

which is greater than the MTR of 1.00E-04 /yr.

Therefore a SIF is required.

PFD = 2.50E-03

Part i)

$$F_{\text{fatality}} = F_{\text{temperature}} \times P_{\text{occupied}} \times P_{\text{mask}} \times P_{\text{control}}$$
$$= 3.60\text{E-}01 \text{ /yr} \times (2\text{hrs} / 24\text{hrs}) \times 0.1 \times 0.1$$
$$= 3.00\text{E-}04 \text{ /yr}$$

which exceeds the MTR for multiple fatalities, 1.00E-05 /yr.

Therefore a SIF is required.

PFD = 3.33E-02

Part j)

$$F_{\text{fatality}} = F_{\text{operation}} \times P_{\text{error}} \times P_{\text{ignition}} \times P_{\text{survival}}$$
$$= 12 \text{ /yr} \times 0.01 \times 0.01 \times 0.5$$
$$= 6.00\text{E-}04 \text{ /yr}$$

which exceeds the MTR for a single fatality, 1.00E-04 /yr.

Therefore a SIF is required.

PFD = 1.67E-01

Example 25, p.29

a) SIL2; b) SIL1; c) SIL2; d) SIL1; e) None; f) SIL1; g) None; h) SIL1.

Example 26, p.31

Part a)

F_{MTR}	=	$F_{control} \times P_{manned}$
1.00E-04 /yr	=	$F_{control} \times$ (2 hr / 24 hr)
$F_{control}$	=	1.00E-04 /yr / 0.083
	=	1.20E-03 /yr
PFH	=	1.37E-07 /hr

Part b)

F_{MTR}	=	$F_{control} \times P_{manned} \times P_{PSV}$
1.00E-04 /yr	=	$F_{control} \times$ (8 hr / 24 hr) $\times$ 0.01
$F_{control}$	=	1.00E-04 /yr / (0.33 $\times$ 0.01)
	=	3.00E-02 /yr
PFH	=	3.42E-06 /hr

Part c)

F_{MTR}	=	$F_{control} \times P_{manned} \times P_{ignition}$
1.00E-04 /yr	=	$F_{control} \times$ (4 hr / 24 hr) $\times$ 0.35
$F_{control}$	=	1.00E-04 /yr / (0.167 $\times$ 0.35)
	=	1.71E-03 /yr
PFH	=	1.96E-07 /hr

10. Answers

Part d)

F_{MTR}	=	$F_{control} \times P_{manned}$
1.00E-05 /yr	=	$F_{control} \times$ (10 mins / 1440 mins)
$F_{control}$	=	1.00E-05 /yr / 6.94E-03
	=	4.11E-03 /yr
PFH	=	4.70E-07 /hr

Part e)

F_{MTR}	=	$F_{control} \times P_{fatality}$
1.00E-05 /yr	=	$F_{control} \times$ (1 / 200)
$F_{control}$	=	1.00E-05 /yr / 5.00E-03
	=	2.00E-03 /yr
PFH	=	2.28E-07 /hr

Part f)

F_{MTR}	=	$F_{control} \times P_{explosion}$
1.00E-04 /yr	=	$F_{control} \times$ 0.1
$F_{control}$	=	1.00E-04 /yr / 0.1
	=	1.00E-03 /yr
PFH	=	1.14E-07 /hr

Part g)

F_{MTR} = $F_{maintenance} \times P_{error} \times P_{ignition}$

1.00E-04 /yr = 1 /yr $\times P_{error} \times$ 0.01

P_{error} = 1.00E-04 /yr / 0.01 /yr

= 1.00E-02 (1% probability of error)

Example 27, p.35

    a) 5.0E+05 hours;

    b) 10 years;

    c) 2220 hours;

    d) 4.55E+05 hours;

    e) 50 years;

    f) 45500 hours;

    g) 1.43E+05 hours;

    h) 10 years.

Example 28, p.36

    a) 100,000 hours;

    b) 62,500 hours;

    c) 454,545 hours;

    d) 18,600 hours;

    e) 12 years.

Example 29, p.36

    a) 4.05×10^{-5} /hr;

    b) 3.81×10^{-5} /hr;

    c) 2.28×10^{-5} /hr;

    d) 8.33×10^{-6} /hr;

    e) 4.08×10^{-4} /hr;

    f) 1.41×10^{-5} /hr;

    g) 1.09×10^{-5} /he;

    h) 2.86×10^{-5} /hr.

Example 30, p.38

    a) 0.9025;

    b) 0.857;

    c) 0.686;

    d) 0.693;

    e) 0.485.

Example 31, p.40

    a) 0.998;

    b) 0.99988;

    c) 0.99997;

    d) 0.997;

    e) 0.9991.

10. Answers

Example 32, p.44

Out of range signal >20mA	DD
Loss of output, signal <4mA	S
Drift and read high	DU
Drift and read low	S

Example 33, p.44

Out of range signal >20mA	S
Loss of output, signal <4mA	DD
Drift and read high	S
Drift and read low	DU

Example 34, p.44

| Stuck closed | DU |
| Spuriously opens | S |

Example 35, p.45

| Fail to close on demand | DU |
| Fail to open on demand | Non-SR |

Example 36, p.45

| Stuck closed | DU |
| Spuriously opens | S |

10. Answers

Example 37, p.45

Coil open circuit	S
Contacts open circuit	S
Contacts short circuit	DU

Example 38, p.45

Fail to close on demand	DU
Leak of the utility medium	S
Internal leakage	DU

Example 39, p.46

Out of range signal >20mA	DD
Loss of output, signal <4mA	S
Drift and read high	DU
Drift and read low	S

Example 40, p.48

$PFD_{DD} = 0.00E+00$

$PFD_{DU} = 2.67E-04$

Example 41, p.49

$PFD_{DD} = 1.94E-06$

$PFD_{DU} = 7.01E-05$

Example 42, p.49

$PFD_{avg} = 7.20E-05$

Example 43, p.49

PFD_{DD} = 1.13E-11

PFD_{DU} = 1.96E-08

Example 44, p.49

PFD_{avg} = 1.97E-08

Example 45, p.51

PFD_{avg} = 3.62E-06

Example 46, p.51

PFD_{avg} = 1.75E-04

Example 47, p.52

    a) 0;

    b) 1;

    c) 1;

    d) 2;

    e) 2.

Example 48, p.53

SFF = 0.58

Example 49, p.53

SFF = 0.90

Example 50, p.53

SFF = 0.80

10. Answers

Example 51, p.54

    a) A;

    b) B;

    c) B;

    d) B;

    e) A;

    f) A;

    g) A;

    h) A.

Example 52, p.57

    a) SIL3;

    b) SIL2;

    c) SIL2;

    d) SIL3;

    e) SIL2;

    f) SIL3;

    g) SIL3;

    h) SIL1.

11. REFERENCES

11.1 Functional Safety in the Process Industry: A handbook of practical guidance in the application of IEC61511 and ANSI/ISA-84.00.01. KJ Kirkcaldy and D Chauhan, ISBN 978-1-291-18723-6.

11.2 IEC61508:2010, Functional Safety of Electrical/ Electronic/ Programmable Electronic Safety Related Systems.

11.3 IEC615112004: Functional Safety: Safety Instrumented Systems for the Process Industry.

12. DEFINITIONS

DEFINITIONS	
2oo3	Two out of three logic circuit (2/3 logic circuit) A logic circuit with three independent inputs. The output of the logic circuit is the same state as any two matching input states. For example a safety circuit where three sensors are present and a signal from any two of those sensors is required to call for a shut down. This 2oo3 system is said to be single fault tolerant (HFT = 1) in that one of the sensors can fail dangerously and the system can still safely shut down. Other voting systems include 1oo1, 1oo2, 2oo2, 1oo3 and 2oo4.
IEC61508	The IEC standard covering Functional Safety of electrical / electronic / programmable electronic safety-related systems The main objective of IEC61508 is to use safety instrumented systems reduce risk to a tolerable level by following the overall, hardware and software safety lifecycle procedures and by maintaining the associated documentation. Issued in 1998 and 2000, it has since come to be used mainly by safety equipment suppliers to show that their equipment is suitable for use in safety integrity level rated systems.
IEC61511	The IEC standard for use of electrical / electronic / programmable electronic safety-related systems in the process industry. Like IEC 61508 it focuses on a set of safety lifecycle processes to manage process risk. It was originally published by the IEC in 2003 and taken up by the US in 2004 as ISA 84.00.01- 2004. Unlike IEC 61508, this standard is targeted toward the process industry users of safety instrumented systems.
ALARP	As low as reasonably practicable. The philosophy of dealing with risks that fall between an upper and lower extreme. The upper extreme is where the risk is so great that it is rejected completely while the lower extreme is where the risk is, or has been made to be, insignificant. This philosophy considers both the costs and benefits of risk reduction to make the risk "as low as reasonably practicable".
Architectural constraints	Limitations that are imposed on the hardware selected to implement a safety-instrumented function, regardless of the performance calculated for a subsystem. Architectural constraints are specified (in IEC61508-2-Table 2 and IEC 61511-Table 5) according to the required SIL of the subsystem, type of components used, and SFF of the subsystem's components. Type A components are simple devices not incorporating microprocessors, and Type B devices are complex devices such as those incorporating microprocessors. See Fault Tolerance.
Architecture	The voting structure of different elements in a safety instrumented function. See Architectural Constraints, Fault Tolerance and 2oo3.

12. Definitions

DEFINITIONS	
Availability	The probability that a device is operating successfully at a given moment in time. This is a measure of the "uptime" and is defined in units of percent. For most tested and repaired safety system components, the availability varies as a saw tooth with time as governed by the proof test and repair cycles. Thus the integrated average availability is used to calculate the average probability of failure on demand. See PFDavg.
Basic process control system	System which responds to input signals from the process, associated equipment, and/or an operator and generates output signals causing the process and its associated equipment to operate in the desired way. The BPCS cannot perform any safety instrumented functions rated with a safety integrity level of 1 or better unless it meets proven in use requirements. See proven in use.
BPCS	See Basic Process Control System.
Cause and effect diagram	One method commonly used to show the relationship between the sensor inputs to a safety function and the required outputs. Often used as part of a safety requirements specification. The method's strengths are a low level of effort and clear visual representation while its weaknesses are a rigid format (some functions cannot be represented w/ C-E diagrams) and the fact that it can oversimplify the function.
Common mode failure	A random stress that causes two or more components to fail at the same time for the same reason. It is different from a systematic failure in that it is random and probabilistic but does not proceed in a fixed, predictable, cause and effect fashion. See systematic failure.
Consequence	The magnitude of harm or measure of the resulting outcome of a harmful event. One of the two components used to define a risk.
Proof test coverage	The percentage failures that are detected during the servicing of equipment. In general it is assumed that when a proof test is performed any errors in the system are detected and corrected (100% proof test coverage).
D Diagnostics	Some safety rated logic solvers are designated as having capital D diagnostics. These are different from regular diagnostics in that the unit is able to reconfigure its architecture after a diagnostic has detected a failure. The greatest effect is for 1oo2D systems which can reconfigure to 1oo1 operation upon detecting a safe failure. Thus the spurious trip rate for such a system is dramatically reduced.
Dangerous failure	A failure of a component in a safety instrumented function that prevents that function from achieving a safe state when it is required to do so. See failure mode.
Diagnostic coverage	A measure of a system's ability to detect failures. This is a ratio between the failure rates for detected failures to the failure rate for all failures in the system.

12. Definitions

DEFINITIONS	
E/E/PE Electrical / Electronic / Programmable Electronic	See 61508 and 61511.
Event tree analysis	A method of fault propagation modelling. The analysis constructs a tree-shaped picture of the chains of events leading from an initiating event to various potential outcomes. The tree expands from the initiating event in branches of intermediate propagating events. Each branch represents a situation where a different outcome is possible. After including all of the appropriate branches, the event tree ends with multiple possible outcomes.
Fail close	A condition wherein the valve closing component moves to a closed position when the actuating energy source fails.
Fail open	A condition wherein the valve closing component moves to an open position when the actuating energy source fails.
Fail safe (or preferably de-energize to trip)	A characteristic of a particular device which causes that device to move to a safe state when it loses electrical or pneumatic energy.
Failure modes	The way that a device fails. These ways are generally grouped into one of four failure modes: Safe Detected (SD), Dangerous Detected (DD), Safe Undetected (SU), and Dangerous Undetected (DU) per ISA TR84.0.02.
Failure rate	The number of failures per unit time for a piece of equipment. Usually assumed to be a constant value. It can be broken down into several categories such as safe and dangerous, detected and undetected, and independent/normal and common cause. Care must be taken to ensure that burn in and wear-out are properly addressed so that the constant failure rate assumption is valid.
Fault tolerance	Ability of a functional unit to continue to perform a required function in the presence of random faults or errors. For example a 1oo2 voting system can tolerate one random component failure and still perform its function. Fault tolerance is one of the specific requirements for safety integrity level (SIL) and is described in more detail in IEC61508 Part 2 Tables 2 and 3 and in IEC61511 (ISA 84.01 2004) in Clause 11.4
Fault tree diagram	Probability combination method for estimating complex probabilities. Since it generally takes the failure view of a system, it is useful in multiple failure mode modelling. Care must be taken when using it to calculate integrated average probabilities.
FMECA	Failure Modes Effects and Criticality Analysis - This is a detailed analysis of the different failure modes and criticality analysis for a piece of equipment.

12. Definitions

DEFINITIONS	
Functional safety	Freedom from unacceptable risk achieved through the safety lifecycle. See IEC61508, IEC65111, safety lifecycle, and tolerable risk.
Hazard	The potential for harm.
HAZOP	Hazards and operability study. A process hazards analysis procedure originally developed by ICI in the 1970s. The method is highly structured and divides the process into different operationally-based nodes and investigates the behaviour of the different parts of each node based on an array of possible deviation conditions or guidewords.
HFT	Hardware fault tolerance (see fault tolerance)
HSE (UK)	Health and Safety Executive
IEC	International Electrotechnical Commission. A worldwide organization for standardization. The object of the IEC is to promote international cooperation on all questions concerning standardization in the electrical and electronic fields. To this end and in addition to other activities, the IEC publishes international standards. See 61508 and 61511. Impact analysis activity of determining the effect that a change to a function or component will have to other functions or components in that system as well as to other systems
Incident	The result of an initiating event that is not stopped from propagating. The incident is most basic description of an unwanted accident, and provides the least information. The term incident is simply used to convey the fact that the process has lost containment of the chemical or other potential energy source. Thus the potential for causing damage has been released but its harmful result has not has not taken specific form.
IPL	Independent protection layer or layers. This refers to various other methods of risk reduction possible for a process. Examples include items such as rupture disks and relief valves which will independently reduce the likelihood of the hazard escalating into a full accident with a harmful outcome. In order to be effective, each layer must specifically prevent the hazard in question from causing harm, act independently of other layers, have a reasonable probability of working, and be able to be audited once the plant is operation relative to its original expected performance.
Lambda	Failure rate for a system. See failure rate.
Likelihood	The frequency of a harmful event often expressed in events per year or events per million hours. One of the two components used to define a risk. Note that this is different from the traditional English definition that means probability.

12. Definitions

DEFINITIONS	
LOPA	Layer of Protection Analysis. A method of analysing the likelihood (frequency) of a harmful outcome event based on an initiating event frequency and on the probability of failure of a series of independent layers of protection capable of preventing the harmful outcome.
Mode (Continuous)	When demands to activate a safety function (SIF) are frequent compared to the test interval of the SIF. Note that other sectors define a separate high demand mode, based on whether diagnostics can reduce the accident rate. In either case, the continuous mode is where the frequency of an unwanted accident is essentially determined by the frequency of a dangerous SIF failure. When the SIF fails, the demand for its action will occur in a much shorter time frame than the function test, so speaking of its failure probability is not meaningful. Essentially all of the dangerous faults of a SIF in continuous mode service will be revealed by a process demand instead of a function test. See low demand mode, high demand mode, and SIL.
Mode (High Demand)	(also continuous mode per IEC61511) Similar to continuous mode only there is specific credit taken for automatic diagnostics. The split between high demand and continuous mode is whether the automatic diagnostics are run many times faster than the demand rate on the safety function. If the diagnostics are slower than this there is no credit for them and the continuous mode applies.
Mode (Low Demand)	(also demand mode per IEC61511) when demands to activate the safety instrumented function (SIF) are infrequent compared to the test interval of the SIF. The process industry defines this mode when the demands to activate the SIF are less than once every two proof test intervals. The low demand mode of operation is the most common mode in the process industries. When defining safety integrity level for the low demand mode, a SIF's performance is measured in terms of average Probability of Failure on Demand (PFDavg). In this demand mode, the frequency of the initiating event, modified by the SIF's probability of failure on demand times the demand rate and any other downstream layers of protection determine the frequency of unwanted accidents.
MTTR	Mean Time to Repair – The average time between the occurrence of a failure and the completion of the repair of that failure. This includes the time needed to detect the failure, initiate the repair and fully complete the repair.
Occupancy	A measure of the probability that the effect zone of an accident will contain one or more personnel receptors of the effect. This probability should be determined using plant-specific staffing philosophy and practice.

12. Definitions

DEFINITIONS	
P&ID	Piping and instrumentation drawing. Shows the interconnection of process equipment and the instrumentation used to control the process. In the process industry, a standard set of symbols is used to prepare drawings of processes. The instrument symbols used in these drawings are generally based on Instrument Society of America (ISA) Standard S5. 1. 2. The primary schematic drawing used for laying out a process control installation.
PFDavg	Probability of Failure on Demand average- This is the probability that a system will fail dangerously, and not be able to perform its safety function when required. PFD can be determined as an average probability or maximum probability over a time period. IEC61508/61511 and ISA 84.01 use PFDavg as the system metric upon which the SIL is defined.
Proof test	Testing of safety system components to detect any failures not detected by automatic on-line diagnostics i.e. dangerous failures, diagnostic failures, parametric failures followed by repair of those failures to an equivalent as new state. Proof testing is a vital part of the safety lifecycle and is critical to ensuring that a system achieves its required safety integrity level throughout the safety lifecycle.
Protection layer	See IPL.
Proven in use	Basis for use of a component or system as part of a safety integrity level (SIL) rated safety instrumented system (SIS) that has not been designed in accordance with IEC61508. It requires sufficient product operational hours, revision history, fault reporting systems, and field failure data to determine if the is evidence of systematic design faults in a product. IEC 61508 provides levels of operational history required for each SIL.
Proof Test Interval	The time interval between servicing of the equipment.
Random failure	A failure occurring at a random time, which results from one or more degradation mechanisms. Random failures can be effectively predicted with statistics and are the basis for the probability of failure on demand based calculations requirements for safety integrity level. See systematic failure.
Redundancy	Use of multiple elements or systems to perform the same function. Redundancy can be implemented by identical elements (identical redundancy) or by diverse elements (diverse redundancy). Redundancy of primarily used to improve reliability or availability.
Reliability	1. The probability that a device will perform its objective adequately, for the period of time specified, under the operating conditions specified. 2. The probability that a component, piece of equipment or system will perform its intended function for a specified period of time, usually operating hours, without requiring corrective maintenance.

12. Definitions

DEFINITIONS	
Reliability block diagram	Probability combination method for estimating complex probabilities. Since it generally takes the "success" view of a system, it can be confusing when used in multiple failure mode modelling.
RRF	Risk Reduction Factor -The inverse of PFDavg
Safe failure	Failure that does not have the potential to put the safety instrumented system in a dangerous or fail-to-function state. The situation when a safety related system or component fails to perform properly in such a way that it calls for the system to be shut down or the safety instrumented function to activate when there is no hazard present.
Safe failure fraction	See SFF.
Safe state	The state of the process after acting to remove the hazard resulting in no significant harm.
SFF	Safe Failure Fraction - The fraction of the overall failure rate of a device that results in either a safe fault or a diagnosed (detected) unsafe fault. The safe failure fraction includes the detectable dangerous failures when those failures are annunciated and procedures for repair or shutdown are in place.
SIF	Safety Instrumented Function – A set of equipment intended to reduce the risk due to a specific hazard (a safety loop). Its purpose is to: 1. Automatically taking an industrial process to a safe state when specified conditions are violated; 2. Permit a process to move forward in a safe manner when specified conditions allow (permissive functions); or 3. Taking action to mitigate the consequences of an industrial hazard. It includes elements that detect an accident is imminent, decide to take action, and then carry out the action needed to bring the process to a safe state. Its ability to detect, decide and act is designated by the safety integrity level (SIL) of the function. See SIL.
SIL	Safety Integrity Level - A quantitative target for measuring the level of performance needed for safety function to achieve a tolerable risk for a process hazard. Defining a target SIL level for the process should be based on the assessment of the likelihood that an incident will occur and the consequences of the incident. The following table describes SIL for different modes of operation.
SIL verification	The process of calculating the average probability of failure on demand (or the probability of failure per hour) and architectural constraints for a safety function design to see if it meets the required SIL.

12. Definitions

DEFINITIONS	
SIS	Safety Instrumented System – Implementation of one or more Safety Instrumented Functions. A SIS is composed of any combination of sensor(s), logic solver(s), and final element(s). A SIS is usually has a number of safety functions with different safety integrity levels (SIL) so it is best avoid describing it by a single SIL. See SIF.
Spurious trip	See Safe failure
Systematic failure	A failure that happens in a deterministic (non random) predictable fashion from a certain cause, which can only be eliminated by a modification of the design or of the manufacturing process, operational procedures, documentation, or other relevant factors. Since these are not mathematically predictable, the safety lifecycle includes a large number of procedures to prevent them from occurring. The procedures are more rigorous for higher safety integrity level systems and components. Such failures cannot be prevented with simple redundancy.

13. INDEX

Architectural Constraints 53
Availability 33
Common Cause Failures 49
Continuous Mode 24
dangerous failure 41
Demand Mode 24, 25
detected failures 42
Failure 33
Failure rate 33, 34
IEC iv
International Electrotechnical Commission iv
Mean Down Time 47
Mean time between failures 33
Parallel Systems 39
Probability of Failure on Demand 23, 47
Proof Test 47
Reliability 33
Reliability Block Diagrams 38
Risk Reduction Factor 23
safe failure 41
Safe Failure Fraction 52
Series Systems 38
SIL Targets 25, 29
undetected failures 42
Value of Preventing a statistical Fatality 18

www.ingramcontent.com/pod-product-compliance
Lightning Source LLC
Chambersburg PA
CBHW072228170526
45158CB00002BA/799